*Geography and the
Environment*

Geography and the Environment

Systems Analytical Methods

A. G. Wilson

School of Geography, University of Leeds
Leeds LS2 9JT, England

JOHN WILEY & SONS

Chichester · New York · Brisbane · Toronto

Copyright © 1981 by John Wiley & Sons Ltd.

All rights reserved.

No part of this book may be reproduced by any means, nor transmitted, nor translated into a machine language without the written permission of the publisher.

British Library Cataloguing in Publication Data:

Wilson, Alan Geoffrey
  Geography and the environment.
  1. Geography—Methodology
  2. Systems theory
  I. Title
  910   G70   80-41696

  ISBN 0 471 27956 0 (cloth)
  ISBN 0 471 27957 9 (paper)

Photosetting by Thomson Press (India) Ltd., New Delhi and Printed in the United States of America.

Contents

PART 2 METHODS OF SYSTEMS THEORY AND SPATIAL ANALYSIS

Chapter 6 Optimization

Chapter 7 Locational and network structures: nodes and links

PART 3   EXAMPLES OF ENVIRONMENTAL SYSTEMS ANALYSIS

Chapter 10   Applications and conclusions

Preface

The 'environment' which forms the subject matter of this book is taken to include natural, man–nature, and manufactured 'systems of interest'. The study of the environment is the concern of a number of disciplines, and although the title identifies geography as the principal one, it is hoped that the book will be of interest to practitioners who see themselves wearing other disciplinary hats—whether in fields like economics, civil engineering, or town planning, or in essentially interdisciplinary groupings like regional science or 'environment studies'.

The 'systems' approach to environmental studies has been fashionable for a number of years and has also been the subject of much controversy about its ultimate usefulness. The basis of the approach lies in a number of concepts, described in Chapter 2, which are intended to deepen understanding of complex systems made up of a large number of interdependent elements. The main argument of this book is that it is necessary to go further: to show how a wider range of methods can be drawn within the 'systems' umbrella. In particular, it is argued that these methods provide the basis for the construction of environmental system models for many examples. This illustrates another feature of the systems analytical approach: not only are there methods available for handling complexity, but these methods can be understood in a general way and it is possible to learn to identify the system 'types' to which they are applicable. Needless to say, there remain many difficult research problems, but it can be argued that sufficient progress has been made to make the presentation of the current state of knowledge available in textbook form.

The book can be used for teaching in a number of ways. It is hoped that it provides an adequate sequence for study as a whole. My own current practice is to teach the *qualitative* content of the book, essentially Chapters 1, 3, and 9, together with an account of how to *describe* environmental systems in mathematical terms and a qualitative description of the methods of Chapters 4–8, in the first year of an undergraduate course, and to teach the methods in more detail in the second year. If the topics are presented in the greatest possible depth, and are supplemented with the literature cited in the bibliography, the book can form the basis of a postgraduate course. It will also be possible for the teacher to provide, or the student to discover, a much wider range of examples which illustrate the general argument and which may be more suitable for particular contexts.

In the Notes on Further Reading at the end of each chapter, references preceded by a dagger are useful complements to the present text at roughly the same level; those preceded by an asterisk are at a more advanced level.

The idea of presenting a series of lectures on methods was first suggested to me by David Boyce and Britton Harris of the University of Pennsylvania, and I am grateful to them and their colleagues for inviting me to lecture there in three successive years from 1976. I am also grateful to my own colleagues and students in the University of Leeds for comments which have arisen from the use of this material in lectures there. Rosanna Whitehead typed some of the early chapters and Pamela Talbot the remainder. Gordon Bryant, Tim Hadwin, and John Dixon drew the figures. I am most grateful to both groups for their crucial work during the preparation of the book.

ALAN WILSON

Leeds
June 1980

Acknowledgements

Figures 1.1, 1.2, and 1.6 are based on Figures 2.2, 5.7, and 4.3 from C. H. Gimingham (1975) *An Introduction to Heathland Ecology*, Oliver and Boyd, Edinburgh.
Figure 1.5 is based on Figure 30 from E. A. FitzPatrick (1974) *An Introduction to Soil Sciences*, Oliver and Boyd, Edinburgh.
All reproduced by permission of Oliver and Boyd, Edinburgh.

Figures 1.3, 2.12, and 2.13 are based on Figures 3.1, 1.2, and 4.6 from E. P. Odum (1975) *Ecology: The Link Between the Natural and Social Sciences*, 2nd edn, Holt, Rinehart and Winston, New York.
All reproduced by permission of Holt, Rinehart and Winston, New York.

Figure 1.4, Copyright © 1975 McGraw-Hill Book Company (UK) Ltd. From Ward *Principles of hydrology, 2nd edn.*
Reproduced by permission.

Figure 2.14 is based on Figure 1.5 from R. C. Ward (1975) *Principles of hydrology*, 2nd edn, McGraw-Hill, U.K.
Reproduced by permission of Dr J. Lewin.

Table 2.1 is reproduced from Table 1.1 of J. G. Miller (1978) *Living Systems*, McGraw-Hill, New York.
Reproduced by permission of McGraw-Hill Book Company, New York.

Figures 1.7, 1.8, 10.5, and 10.8 are based on Figures 17.1, 17.2, 13.8, and 18.1 from R. de Neufville and D. H. Marks (eds.) (1974) *Systems Planning and Design*, Prentice-Hall, Englewood Cliffs.
All reproduced by permission of Prentice-Hall Inc., Englewood Cliffs, N. J., and the authors.

Figures 1.11, 2.11, and 2.19 are based on Figures 8.21 and 1.2 from R. J. Chorley and B. A. Kennedy (1971) *Physical Geography: A Systems Approach*, Prentice-Hall, London.
All reproduced by permission of the authors.

Figures 1.9 and 10.9 are based on Figure 18 from A. G. Wilson (1972) Understanding the city of the future. *University of Leeds Review*, 15.
Reproduced by permission of the University of Leeds.

Figure 1.10 is based on Figure 1 (p. 37) from P. Cowan, *et al.* (1973) *The Future of Planning*, Heinemann, London.
Reproduced by permission of Heinemann, London.

Figure 2.1 is based on the figure on p. 114 of S. Beer (1967) *Management Science*, Aldus Books, London.
Reproduced by permission of Aldus Books, London.

Figure 2.8 is based on Figure 1.4(a) from J. W. Forrester (1968) *Principles of Systems*, Wright Allen Press, Mass.
Reproduced by permission of MIT Press, Cambridge, Mass.

Figures 2.9 and 2.10 are based on Figures 2.5, 2.6, 2.7, 2.8, and 2.9 from J. Klir and M. Valach (1967) *Cybernetic Modelling*, Iliffe Books, London.
All reproduced by permission of Butterworths, Sevenoaks.

Figures 2.15, 2.16 and 8.6 are based on Figures 2.4, 2.5, and A-1 from J. W. Forrester (1969) *Urban Dynamics*, MIT Press, Mass.
All reproduced by permission of MIT Press, Cambridge, Mass.

Figure 7.8 is based on Figure 4.14 from F. S. Roberts (1976) *Discrete Mathematical Models*, Prentice-Hall, Englewood Cliffs.
Reproduced by permission of the U.S. Army Corps of Engineers.

Figure 7.9 is based on Figure 1 from R. H. Atkin (1977) *Combinatorial Connectivities in Social Systems*, Berkhauser Verlag, Basel.
Reproduced by permission of Birkhauser Verlag, Basel.

Figure 7.26 is based on Figure 4.6 from F. S. Roberts (1976) *Discrete Mathematical Models*, Prentice-Hall, Englewood Cliffs.
Reproduced by permission of American Scientist, Journal of Sigma Xi, The Scientific Research Society.

Figure 7.27 is based on Figure 4.8 from F. S. Roberts (1976) *Discrete Mathematical Models*, Prentice-Hall, Englewood Cliffs.
Reproduced by permission of Pion Ltd., London.

Figure 7.28 is based on Figure 4.50 from F. S. Roberts (1976) *Discrete Mathematical Models*, Prentice-Hall, Englewood Cliffs.
Reproduced by permission of the Rand Corporation, Santa Monica.

Figure 8.8 is based on Figure C (p. 268) from M. W. Hirsch and S. Smale (1974) *Differential Equations, Dynamical Systems and Linear Algebra*, Academic Press, New York.
Reproduced by permission of Academic Press Inc., New York, and the authors.

Figures 8.11, 8.12, 8.13, 8.14, 8.15, 8.16, and 8.17 are based on Figure 1 from A. G. Wilson (1976) Catastrophe theory and urban modelling: an application to modal choice, *Environment and planning A*, **8**, 351–6; Figure 8 from J. C. Amson (1975) Catastrophe theory: a contribution to the study of urban systems, *Environment and Planning B*, **2**, 177–221; and Figures 1, 4, 5, and 6 from T. Poston and A. G. Wilson (1977) Facility size versus distance travelled: urban services and the fold catastrophe, *Environment and Planning A*, **9**, 681–6.
All reproduced by permission of Pion, Ltd., London, and Dr J. C. Amson (for Figures 8.12 and 8.13).

Figures 9.1 and 9.2 are based on Figures 22 and 27 from S. Beer (1972) *Brain of the Firm*, Allen Lane, London.
Both reproduced by permission of the author.

Figure 9.3 is based on Figure 2.4 from A. G. Wilson (1974) *Urban and Regional Models in Geography and Planning*, John Wiley, Chichester.
Reproduced by permission of John Wiley & Sons, Chichester.

Figure 9.4 is based on Figure 9.2 from J. B. McLoughlin (1977) *Control And Urban Planning*, Faber and Faber, London.
Reproduced by permission of Faber and Faber, London.

Figures 10.3 and 10.4 are based on Tables 1.1 and 1.2 from Yorkshire and Humberside Economic Planning Board (1976) *The Pennine Uplands*, HMSO, London.
Reproduced by permission of the Controller of Her Majesty's Stationery Office.

PART 1

Basic Ideas of Systems Analysis

CHAPTER 1

An informal introduction to systems analysis

1.1 AIMS

Geography, planning, and the environmental sciences generally are among the many disciplines which have usefully incorporated concepts of systems analysis (or, more broadly, systems *theory*). The concept of a system will be refined and elaborated at various stages in this book, but it can be defined initially simply as an object of study which is made up of a number (usually large) of interrelated components. Many, if not most, objects of study are defined as systems on this basis, and hence the widely pervading influence of systems analysis. The usual emphases of a systems-analytical study, however, restrict the field of application to some extent: the main concern is with *complicated* systems whose components exhibit high degrees of *interdependence*. The behaviour of the 'whole' system is then usually something very much more than the sum of the parts.

The arguments of systems theory are sometimes clouded by jargon, and the primary aim of this book is to present the main concepts and methods, as they relate to the environmental sciences, as clearly as possible. Technical terms will be introduced where necessary, but defined where possible from first principles. A subsidiary but important aim of the book will be a discussion of how other authors use systems concepts (since frequently the same words are used in different ways by different people) to help to provide a guide to the wider literature. It is hoped that the concepts and methods of systems theory will be absorbed by the reader in a natural way, and that they can then be applied within a wide variety of situations. A number of examples are used to illustrate the argument at various stages but these can be added to substantially from the literature. One of the main canons of systems theory is that common methods of analysis can often be applied to wide varieties of apparently very different systems. The reader should seek to acquire the ability to transfer methods to new situations, though not, of course, without exercising great caution. In many cases, particularity and uniqueness are more important than generality, and judgements have to be made about this.

In the remaining three subsections of this introductory chapter, a range of preliminary ground is covered. In Section 1.2, the broad nature of systems

3

analysis and theory is explored; in Section 1.3 the place of systems theory within science (and arts) is reviewed; and in Section 1.4, three examples are presented — of moorland ecosystems, water resource systems and cities — to illustrate the need for a systems approach. The formal bases of systems theory are then presented in Chapter 2 with a discussion of concepts, diagrams, types of problem, types of system, models, and theories about systems. This completes the presentation of basic introductory ideas in Part 1.

A 'methods' perspective is used for Part 2, which is made up of Chapters 3 to 9. Methods of progressively increasing complexity are described and related to the discussion of system types in Chapter 2. The aim is to present these discussions as untechnically as possible, and in particular to minimize demands on advanced mathematical techniques (though a mathematical description is often appropriate). A broad conceptual introduction is thus provided, and the reader who wishes to pursue particular topics further is offered a guide into the literature.

In Part 3 (Chapter 10), it is emphasized that an eclectic synthesis of methods is usually needed in any particular case, and some examples of this approach are given.

1.2 ON UNDERSTANDING AND PLANNING COMPLICATED SYSTEMS

The basic aim of systems theory is to generate a deeper understanding of objects of study — systems of interest. This understanding can then often be deployed in the planning of such systems in the future. The procedures we use for gaining understanding can be called analysis or theory building, and hence the use of 'systems analysis' and 'systems theory' almost interchangeably. Analysis has connotations of general method; while theory suggests something slightly more specific, applicable perhaps to a particular type of system. The concept of theory will be treated more formally in the next section.

In seeking to understand complicated systems, three main objectives can always be borne in mind. First, we seek to *set up frameworks to handle complexity*. Initially, this can be a concern with the *systematic description* of the object of study and its associated *structure* (its components) and *processes* (ways in which parts or the whole change). Secondly, we may seek to identify genuinely *systemic behaviour*, where the behaviour of the whole system is in some surprising way greater than would be predicted as the sum of its parts. Such behaviour is a likely consequence of a high level of interdependence. Thirdly, in studying particular systems in this way, we seek *generality*: whether methods of analysis devised for one system can be applied to others; whether more general insights can be gained.

It is important to note that the first of these objectives stands independently of the second; and the first two independently of the third. As will be argued later in this chapter, the first objective is little more than a restatement of one of the standard objectives of science, but it does emphasize complexity together with, in recognizing interdependence, a need to attempt to study 'whole' systems. This is in itself important, though, again as we shall see later, the definition of system boundaries is often a matter for judgement.

1.3 THE CONTEXT: SYSTEMS ANALYSIS WITHIN SCIENCE

1.3.1 The main concepts of scientific method

It was Sir Peter Medawar (1969) who remarked that science had undoubtedly been one of the most successful enterprises ever undertaken but that it would be difficult to get a scientist to give a reasonable account of his underlying methods. Indeed, it could be argued that there is so much confusion and misconception about the nature of scientific method that it is essential to try to clarify some of the associated concepts before tackling something like systems theory. It is often suggested, for example, that the scientist derives general truths from facts and that the scientific method provides an infallible way of doing this. Even concepts like 'truth' and 'facts' are often dangerous oversimplifications. In a sense, there is no such thing as a fact—only observations perceived by particular people in some way at some time. Equally, things which are believed to be true at one time may not be believed to be true at another—and this is as true of the subject matter of science as, say, politics. The view of the scientist as someone discovering general truths from facts is essentially Baconian. It has only been in this century that philosophers like Sir Karl Popper (1959) have recognized that whatever the attractions of such an ideal, it certainly does not represent the way the successful scientist works.

It can be argued that the essence of the scientific method is the construction of *theories* and the continual testing of these by comparing predictions from them with observations. The essence of such testing is an attempt to disprove the theory—to marshal observations which contradict the predictions of the theory. In other words, Popper has argued that the scientist is concerned with attempted *falsification* rather than verification directly. In this sense, theories are never proved to be generally true but only 'not yet proved false'. They start out life as *hypotheses*: a set of propositions which aim to explain the structure of some system and how it develops. That is, we are interested in *pattern* (or structure) and *process*. The key term in this discussion of theories and hypotheses is 'explain'. The propositions may be expressed in a variety of languages. In the case of some complicated systems, explanations can only be adequately achieved with the help of mathematical language. It should also be noted that explanation can be at various levels and that the scientist is continually seeking more depth. Systematic description, for example, may be seen as a useful, but essentially very low, level of explanation. A high level of explanation is often a theory which contains abstract concepts. It is also interesting to note that, in Popper's terms, the higher the level of explanation, the more is being asserted about the world, and therefore such a theory is potentially the more falsifiable.

A *theory* is usually used to denote a hypothesis which has been tested to a reasonable extent and which can be believed with more confidence—though note the subjective features associated with 'believe' and 'confidence'. Although it has been formally established since the work of David Hume in the 18th century that theories can never be fully verified and shown to be generally true (the so-called problem of induction), it is useful to introduce the concept of a *law* as a very well tested and established theory. It is important, however, always to remember that even the best laws are likely to be revised and replaced in due course — such as the replacement of Newton's classical physics by Einstein's relativistic physics.

The main task of the scientist then is to generate and to test theories. There are two fundamentally different approaches to this: *inductive* and *deductive*. The inductive method is usually considered to be the most pure: theoretical preconceptions are to be avoided; the scientist studies his data and aims to infer more general theories. The deductivist on the other hand *speculates* (admittedly, at least usually, in an informed way) on how his system of interest works and builds his theory accordingly. He then makes deductions from his theory which can be tested against observation, and rejects or revises or expands it according to the outcome. The inductive method is essentially the Baconian method mentioned earlier and has come to dominate social science. In more developed sciences, it is customary to pay lip-service to it and then to practise something else. The deductive methods have proved very much more successful. There will be an emphasis here, therefore, on the application of deductive methods in environmental science. However, formal systems theory can be seen as representing another kind of inductive approach to theory building and this idea will be pursued later.

1.3.2 The need for a philosophical basis

It is often said, and with some justice, that it is difficult to find philosophers of science who have ever been original scientists. And we also remarked earlier that some scientists would find it difficult to give an orderly account of their methods. Nonetheless, it is important for any individual to have a clear idea as to how he is employing often used but often less well understood concepts such as *hypothesis*, *theory*, or *law*. His own definitions could then be related, if necessary, to other people's possibly slightly different ones. In other words, it is important to establish the maximum degree of clarity to avoid muddle. There is another equally important reason for attempting this. If accepted views of scientific method are taken uncritically, then the student or researcher is likely to assume that they are the only possible ones. It is this kind of attitude which has led to the dominance of inductive and Baconian methods in social science. So a critical understanding of the variety of methods available is more likely to give the scientist a chance of assembling eclectically a set of methods suitable for a particular problem.

A further and complementary view of science is given by Thomas Kuhn (1962), first presented in his book *The Structure of Scientific Revolutions*. This connects in a way to Ziman's (1968) view of science as a wide-ranging public activity: at any one time, Kuhn argues, there is a consensus about the existing body of knowledge within a science — the current paradigm. 'Normal science' is problem-solving within this paradigm. A situation is then reached where it is recognized that there are problems which cannot be solved within the paradigm. At this stage, major theoretical innovation is needed and perhaps achieved. Kuhn, arguing on the basis of historical evidence, notes that such innovations usually take the form of a revolution and the almost total overthrow of the earlier paradigm.

This leads via Kuhn (1963) to another interesting comment on the nature of scientific discovery. He notes that within a science there are essentially *convergent* phases, where the paradigm is pretty solid, everyone knows what to teach, and most problems appear solvable within the paradigm; and divergent phases, which are more loosely structured and which are problem oriented. First thoughts would suggest that the revolutions are likely to occur during divergent phases, but it is here that Kuhn produces a striking conclusion: historically, the great advances in science are generally made during convergent periods, because only then can the achievement in solving particular problems outside a strong paradigm be recognized as important.

It is interesting, of course, to speculate on the stage of evolution of environmental science. Many of its components, especially on the physical side, such as aspects of ecology or climatology, are parts of 'established' science and associated methods are 'normal-within-a-paradigm'. The normal paradigm on the social science side is less well established, but the foundations are perhaps now apparent. For 'whole' environmental systems, the kinds of methods presented in this book are beginning to lay the foundations of a paradigm. In summary, then, a preliminary view may be that the environmental sciences may be moving from a divergent phase into a more convergent one; and that the resulting normal paradigm would then provide the framework for further revolutions. A more considered view is a matter for the reader on the evidence of the rest of this book and the wider literature!

1.3.3 The relationship to 'arts' approaches

After starting a book on systems theory with a discussion of scientific method, it would seem trite and obvious to state that the whole approach is intended to be scientific. However, there is also an important sense in which this is not intended to be the case. It is possibly especially important to have a brief discussion about the relationship between science and arts approaches in the context of fields such as geography, which would be considered by many to be an arts subject. The argument on this issue which is implicit below can be stated broadly as follows. The arts person and the scientist are each trying to establish, at as general a level as possible, knowledge about their systems of interest. It is obvious to remark that certain kinds of knowledge or insight, into the workings of a city and its social system for example, could be best presented by, and acquired from, a novelist rather than a mathematical modeller. (See Hudson, 1980, for a good statement of this argument in relation to psychology.) In this sense, their roles are com-

8

plementary, but in a philosophical sense not fundamentally different. There are separate arguments, of course, about aesthetic attributes of the arts, but it is convenient for present purposes to settle for a 'meaning theory of art' rather than an aesthetic one.

The similarities can also be emphasized in relation to styles of work. The popular image of the novelist or painter is of the creative and imaginative person; and of the scientist as the 'desiccated calculating machine'. It will be seen by now that this view can only be consistently held by those who take a Baconian view of science and, perhaps, a Bohemian view of art. In reality, the creation and imaginative powers of the great scientist are on a par with those of the great novelist.

Thus, if with systems theory we are in business to achieve the deepest insights on how a great variety of systems of interest work, including socio-economic systems as well as natural systems, then the theory has something to say about the whole spectrum of possible approaches, whether called arts or science. Broadly speaking, therefore, we take a one culture view of arts and science (cf. the 'two cultures' argument between Leavis, 1962, and Snow, 1959).

1.3.4 Systems analysis in this framework

Although the formal definitions are not introduced until the next section, already it has been necessary to use informally a concept of 'system of interest'. At this stage this means no more than the object or set of objects being studied. Systems theory, however, has generated major new fields of research for all kinds of scientists. The traditional system of interest for the scientist could be described by relatively few characteristics, and methods usually involved 'varying one factor at a time'. Such systems of course have their place within systems theory. We will be more concerned though with systems of interest containing large numbers of interacting components. These may be natural ecosystems, resource systems in which man and ecosystems interact, or socio-economic systems such as cities. It can be argued that the *traditional* approach of the scientist in this situation is essentially reductionist: that is, to investigate in the greatest detail the component parts of the system of interest and then to assume that the total assembly of this knowledge will provide the basis for solving any problem relating to that system. The systems theorist, however, argues that the behaviour of the overall system has a distinct, interesting, and often important character of its own. In very many cases, the behaviour of the whole is very much more than the sum of the behaviour of the parts. Human beings provide one very striking example of this: the brain consists of approximately 10^{10} cells, or neurons, and much is known about how particular cells function. But the connections from this sum of parts to the richness and variety of thought generated by the human brain are poorly understood. This is a good example of holistic behaviour presenting new and difficult scientific problems. It will be seen from the examples below that there are similar phenomena in many branches of environmental science. Essentially, complexity and interdependence are seen to generate new kinds of scientific problem, and these new problems constitute much of the subject matter of systems theory. Many traditional approaches to such examples are not only inductive but reductionist also. The interesting questions, therefore, to be explored below are concerned with the need for new methods for studying complex systems and the extent to which we can identify behaviour of whole systems of interest which is distinctive and arises from interdependence and wholeness. Such systems should be related to their component parts as the brain to its neurons.

1.4 THREE EXAMPLES

1.4.1 Introduction

The three examples presented in this section provide illustrations of the concepts introduced informally above. The reader should bear in mind the notions of *complexity* and *interdependence* in relation to the structure and processes associated with each example. A brief systematic description will be given in each case and the advantages to be gained from a comprehensive, non-reductionist, approach will be seen. We will note aspects of systemic behaviour where possible and attempt to sketch the level of understanding and remaining theoretical problems in each case. Planning issues can also be illustrated.

The examples are intended to cover the range which is familiar to the environmental scientist, from the mainly physical (the ecosystem), through an example where man and nature have roughly equal weight (the water resource system), to mainly human and manufactured technology (the city). The purpose of this section is to provide examples which the reader can use to add flesh to the bones of the more abstract concepts to be introduced from Chapter 2 onwards (though further examples will also be introduced in appropriate places). It is *not* the intention to provide a detailed account of work in these areas; the reader is referred to the texts listed at the end of the chapter for such purposes, though some of these texts may be read in a new light using the concepts of systems theory introduced here.

1.4.2 A moorland ecosystem

A moorland provides a good example of an ecosystem for present purposes for a number of reasons: it is simpler (though not simple) than many others; it exhibits all the main features of the argument so far; and it will be familiar to most readers. Systematic description involves enumerating the main components and the interactions between them. This can conveniently be accomplished with two figures from the work of Gimingham (1975), presented here as Figures 1.1 and 1.2. The first of these shows the main plant species and the second the possible animal populations, with some indication of the interrelationships between them. This, therefore, begins to articulate what we mean by interdependence for this example.

The complexity of the system is at once apparent. Within an environment

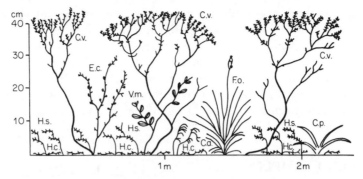

Figure 1.1 Moorland plant species (Gimingham, 1975). Diagram of a section or profile through a heath community (heather with bell heather and blaeberry, N. E. Scotland). C.a.–*Cladonia arbuscula* (lichen); C.p.–*Carex pilulifera* (pill-headed sedge); C.v.–*Calluna vulgaris* (heather); E.c. –*Erica cinerea* (bell heather); F.o.–*Festuca ovina* (sheep's fescue); H.c.!–*Hypnum cupressiforme* (moss); H.s.–*Hylocomium splendens* (moss); V.m. – *Vaccinium myrtillus* (blaeberry)

provided by the climate and the soil, the many plant and animal species will support and compete with each other in a variety of ways. Some of the interaction is direct: animals eating plants, or some animals eating other animals. Some is indirect: different species of plant competing for a common supply of sunlight, or different species of animals competing for a common plant food supply. The potential merits of a comprehensive approach to the analysis of such a system are clear. Any attempt to focus on one species and components immediately connected to it is at best partial though a good approximation, and at worst misleading because of the neglect of what were seen as disconnected effects.

The next step in the argument is to examine the processes involved in system change. The main agent of change and system development in this case is the

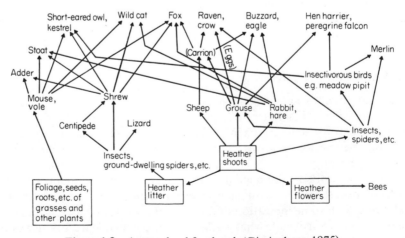

Figure 1.2 A moorland food-web (Gimingham, 1975)

A PICTORIAL DIAGRAM

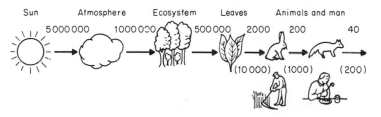

B ENERGY FLOW DIAGRAM

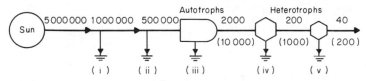

Figure 1.3 Energy flows in ecosystems (Odum, 1975). Solar-energy flow in kcal/m^2 per year, as shown in a pictorial diagram (A) and in a more formalized flow diagram (B). The heat sink symbol ($\doteqdot$) in (B) shows where energy is lost in transformation. Five losses where useful work is done are as follows: (i) Attenuation of extraterrestrial sun energy in heating the atmosphere and driving hydrological cycles and whether systems. (ii) Attenuation of sun energy to warm the ecosystem and drive its internal water and mineral cycles. (iii) Energy loss in conversion of sun energy to plant matter. (iv) Energy loss in conversion of plants to herbivores (primary consumers (C_1)). (v) Energy loss in transfer from primary to secondary consumers (C_2). The figures in parentheses in the biological part of the energy chain represent levels for subsidized ecosystems

supply of light energy from the sun. This is 'fixed' by plants using the process of photosynthesis and passed up through the type of food-web illustrated in Figure 1.2. The types of energy flows involved are shown explicitly in Figure 1.3, using some work by Odum (1975).

This energy also drives the main cycles affecting the ecosystem—the flow of water in the hydrological cycle and various materials' cycles. The hydrological cycle is illustrated in Figure 1.4 and the nitrogen cycle in Figure 1.5. These show important processes taking place in the 'environment' of the ecosystem as defined—the atmosphere and the soil. These processes are themselves interdependent: for example, atmospheric processes such as wind and rain cause certain kinds of soil erosion.

The third kind of process involved is human interference. The use of the moorland for sheep farming or grouse shooting are obvious examples, but in some cases urban pollution may also be relevant. Man, as we will see shortly, can also affect and partially control the system directly by such means as the burning of heather.

It can easily be seen that the system is a complicated one. The processes involved in each component type—plant or animal—will have their own

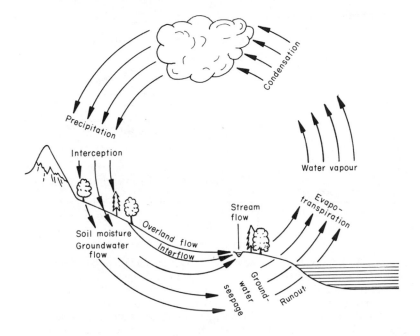

Figure 1.4   The hydrological cycle (Ward, 1973)

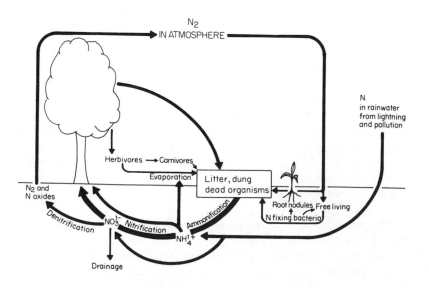

Figure 1.5   The nitrogen cycle (Fitzpatrick, 1974)

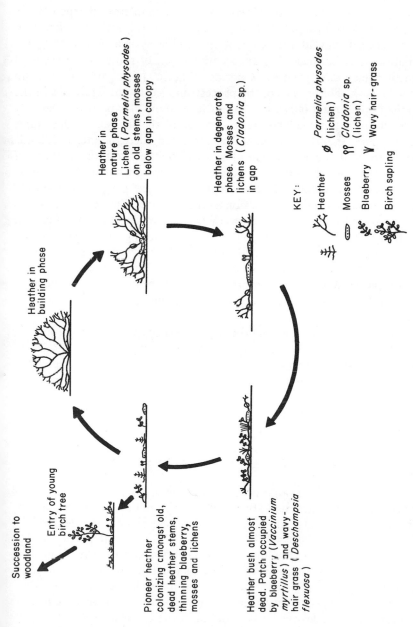

Figure 1.6 The development of an ecosystem—an example (Gimingham, 1975). Cyclical change in a heath community associated with the sequence of growth-phases of heather. Top left: the entry of birch initiates succession towards a woodland community

characteristics, and detailed knowledge of the whole system would demand knowledge of these processes in the context of the energy flows, cycles, and other system-wide processes outlined above. Any attempts to model the whole system and the way it changes should reflect this complexity and interdependence. While a very great deal is known about all these processes, relatively little work has been done in a detailed way at the scale of the whole system. However, some approximations can be made and an account given of some of the types of development which are possible. Consider Figure 1.6, again taken from Gimingham's work. The essence of the process implied by the figure lies in the nature of the development of heather. When very young, it captures a relatively small proportion of the available light (top left of the figure). At this stage, competition is most evenly balanced and, as indicated, succession to woodlands is feasible. In the building and mature phases, the heather 'captures' so much of the light that there is relatively little competition and other plant life will be represented relatively weakly. The ecosystem in this case progresses smoothly towards its climax and possible succession to another state. The actual transition cannot be predicted with absolute accuracy because of the elements of chance involved. It is easy to see how man, via burning or sheep grazing, can interfere with this picture. Typically, the knowledge of past experience is used to plan burning and grazing policies and this may often suffice for the management of a relatively simple ecosystem; but in more complicated cases, an explicit detailed model would be of great advantage in the planning of, say, agricultural development.

It can now be seen that the moorland ecosystem will, typically, have properties at the 'whole system' scale and that there are many possible modes of development. In particular, the varied spatial patterns of 'stands' of vegetation evident on many single moors (say bracker, heather, bilberry, or grass dominated) can be seen to arise from the interplay of phenomena at the holistic scale. The possibilities may not be too surprising, or counter-intuitive, in this case, because much knowledge has been accumulated on this kind of system in the past.

1.4.3 Water resources

Many water systems are almost purely natural ones. In a society with a low level of technological development and abundant water, supplies can be obtained by simple means. As the level of development increases, so does the demand for water for agricultural, industrial, domestic, and recreational purposes. The demand then exceeds low-technology supply; facilities have to be developed for treatment, extraction, distribution, and, perhaps most important of all in scale, storage. In this example, therefore, man and nature each play major roles.

The natural basis for systematic description of a water resource system is the drainage basin, an example of which is shown in Figure 1.7. Water circulates on the basis of the hydrological cycle illustrated earlier and is carried down the basin in a network of channels. An important part of system description is therefore an account of drainage basin morphology, and flow patterns, which are parts of

Figure 1.7 A river basin (Jacoby and Loucks, 1974)

geomorphology and hydrology. These channels carry not only water, but also other inputs such as sewage and other waste and they provide a habitat for fish, animals, and various forms of vegetation. The manufactured elements of such a system include storage facilities (reservoirs), waste emission points, treatment facilities, extraction points and distribution networks, and possibly power plants (hydro-electric facilities). A schematic representation of the river basin shown earlier, displaying the reservoir locations, power plants, and main extraction points is shown in Figure 1.8.

The hydrological cycle has already been mentioned as the driving process of this system. There are also important chemical and biological processes, for example concerned with the breakdown of pollutants. The manufacturing and domestic processes which contribute to waste are of many kinds. Most of these processes are understood individually, but as in the case of the moorland ecosystem, this is less true for the overall effects, such as the density of particular pollutants in the lower reaches of a basin. This, therefore, represents a kind of systemic behaviour. The interdependence in this case arises from the 'connected-ness' of the water throughout the network. At an extraction point, a unit of water may potentially and in practice have been supplied from one of many reservoirs, say, and this could only be identified on a probabilistic basis. This is even more likely to be the case for the distribution of pollutants in a network. And since the source cannot always be identified clearly, this makes the task of pollution control that much more difficult.

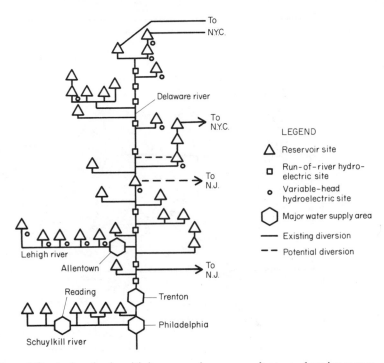

To
N.Y.C.

Delaware river

To
N.Y.C.

To
N.J.

Lehigh river

Allentown

To
N.J.

Reading

Trenton

Philadelphia

Schuylkill river

LEGEND

△ Reservoir site

▫ Run-of-river hydro-
electric site

∘ Variable-head
hydroelectric site

⬡ Major water supply area

— Existing diversion

-- Potential diversion

Figure 1.8 A river basin with its reservoirs, power plants, and major extraction points (Jacoby and Loucks, 1974)

Other kinds of systemic behaviour are more subtle. For example, controls on
water levels in particular reservoirs may be used for flood control; one region of a
system may then 'help' another which does not at first sight appear to be strongly
connected to it. A further example arises when a water-supplying authority has to
meet a particular pattern of demand and it may attempt to do this at minimum
cost subject to constraints on both supply (say, in the reservoirs) and demand
(say, in towns). Such a procedure would associate 'shadow prices' with each
reservoir, which would indicate its role and efficiency in the system as a whole;
and again, factors may influence such prices at a particular place which seem very
remote from that place. This will be pursued further in a later chapter. Finally,
control of reservoir levels for a variety of purposes will affect flows and this will
affect power generating capabilities at different times.

The major planning problems associated with water resource systems are
usually concerned with the siting of new facilities such as reservoirs to meet an
apparently ever-increasing demand. The interconnectedness of the whole system
complicates such decisions enormously by extending the range of possible sites.
There is also a related range of management problems, noted in the previous
paragraph as being interdependent, associated with supply, irrigation, flood
control, and power, for example.

1.4.4 Cities

Cities provide a good illustration both because they are rich in systemic behaviour and because most people come into contact with them so that they function as permanent, if ever-changing, field laboratories. There is a tremendous leap in order of complexity from the two previous examples if only because of the variety of individual lifestyles, realized and potential, of a city's inhabitants. It is useful at the outset to make a distinction between individuals functioning domestically or as consumers, and acting in organizations. The structure of the city can then be seen in terms of population activities (housing, jobs, various services) and organizational, or more usually (if more narrowly), economic, activities together with physical infrastructure and the transport system which support these. Such a structure is exhibited in Figure 1.9.

The driving processes of this kind of structure are birth, death, and migration on the population side and, on the other side, economic development involving changing sectoral mixes and new spatial patterns as a result of technological advance, increasing incomes, and so on. Since the 1950's this has involved decentralization on a substantial scale, increasing per capita incomes and car ownership, increasing travel but declining public transport usage, and, in many places, an increasing spatial and economic polarisation between more- and less-well-off. This is now evident as the much-discussed 'inner-city' problem. Structural interdependences, for example between sectors of the economy, are often very strong and this leads to a variety of systemic effects. Many individual decisions, such as those involving housing and transport, become mutually reinforcing. If a few well off people move from inner city to suburbs, the environment may be perceived as worse and deteriorating for the remainder, and they will be yet more inclined to move out when they can afford it. As travellers switch to car, public transport revenues go down, fares go up, and thus further increase the incentive for more car travel. These, and many other examples, will be pursued as they arise in subsequent chapters.

As might be expected, there is a greater variety of planning problems than for our previous examples, and they are often of greater complexity because of the high levels of interdependence between urban phenomena. Figure 1.10, for

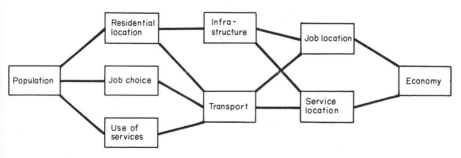

Figure 1.9 The main components of an urban system (Wilson 1972b)

18

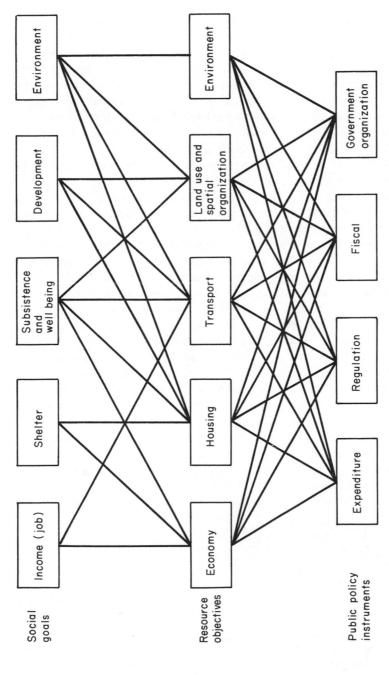

Figure 1.10 Linkages between policy instruments, resource objectives, and social goals (Wilson 1973a)

example, is more-or-less self-explanatory, and shows the relationship between the public policy instruments of urban government attempting to arrange resources optimally to achieve social goals under various heads. The linkages show both high levels of interdependence and 'network' effects, which make it difficult to trace the precise impacts of particular policies. This makes the planning task correspondingly difficult.

1.4.5 Some general points arising from the examples

Three more general points can usefully be noted at this stage. First, by taking a comprehensive system perspective, a viewpoint is established which differs from that of any traditional discipline—except insofar as geography and ecology are themselves synthesizing disciplines. Knowledge has to be brought to bear from many disciplines and subdisciplines to deal systematically with the examples, ranging from biology, climatology, pedology, geomorphology, hydrology, and chemistry to economics, sociology, and civil engineering. An important advantage of a systems approach is the natural tendency to avoid disciplinary myopia.

Secondly, we can begin to see how to achieve greater generality from the study of particular examples and it is worthwhile at this stage to give a foretaste of what is to come. In the moorland ecosystem case, by classifying the components more generally, some of the main interactions can be exhibited in a form which is then applicable to other types of ecosystem. This is illustrated in Figure 1.11, taken from the work of Chorley and Kennedy (1971). The autotrophs are the photosynthesizing plants and the heterotrophs consume these (herbivores) or other heterotrophs (carnivores). Otherwise, the diagram is largely self-explanatory. We will return to other general features of it in the subsection on

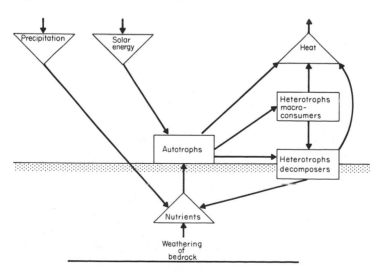

Figure 1.11   The major linkages between the components of a simple terrestrial ecosystem (Chorley and Kennedy, 1971)

system diagrams in Chapter 2. Other more general characteristics emerge in the various 'methods' chapters in Part 2 and further discussion will be postponed until then.

Thirdly, even at this early and informal stage of the argument, it will already be clear that phenomena concerned with whole systems are complicated and that solutions to planning problems should not be simplistic and fail to take this into account. One example will suffice to make the point at this stage. If it is desired to reverse the decline in public transport usage, this almost certainly cannot be achieved purely using instruments of transport policy. For example, research has shown that the introduction of *free* public transport would have a relatively minor impact in this respect. The regulation of car flows on highways, or the imposition of central area car parking charges, would have to be very stringent in order to accomplish a large switch to public transport; and then the secondary effects may become dominant: car drivers may seek new 'centres' away from traditional ones and this may accelerate the decentralization of employment. Figure 1.10 and the preceding discussion suggests that such a problem may at worst be intractable, and at best only solvable in terms of *land use* policy.

NOTES ON FURTHER READING

A useful general introduction on modelling and systems analysis in geography is offered by †Haggett (1965, pp. 17–25), though the treatment is modified in the second edition (†Haggett, Cliff, and Frey, 1977, Chapter 1). A good account of the philosophical notions discussed in Section 1.3 is to be found in †Medawar (1969). Some of the concepts are surveyed in a geographical context in †Wilson (1972a). The three examples were presented in Section 1.4 and it is essential that students who are not familiar with them should undertake some background reading. There are many possible references and other relevant course material would be useful. The books cited here are mainly the direct background to the present text. The short, clear account of the ecology of heathlands is presented by †Gimingham (1975) and in more detail in *Gimingham (1972). A number of useful examples can be culled from the broader treatment of †Odum (1975). Useful sources on water resource systems are †Biswas (1976, especially Chapters 1, 5, 6, 9, and 11) and †de Neufville and Marks (1974, especially Chapters 4, 13, 17, and 18). Each of these books involve more detail than is necessary for the first two chapters of this book, but are all valuable in relation to later chapters (and will also be cited separately in later 'Notes on further reading'). Broad-ranging papers on urban systems are Wilson (1972b, 1973a) with a more specific treatment in †Wilson (1974, Chapter 2). Any of a number of urban geography texts can be consulted, of which a good recent example is †King and Golledge (1978). Broad systems analytical treatments are offered by †Stearns and Montag (1975) and †Vester (1976)—the later could be read in conjunction with the programmed version of its environmental simulation game in †Clarke (1978).

CHAPTER 2

The formal bases of systems analysis

2.1 BASIC CONCEPTS: DEFINITIONS AND DISCUSSION

This is the appropriate point to introduce the basic concepts of systems analysis more formally. The aim is to give definitions which are as simple as possible and which provide a frame of reference not only for the rest of this book, but also for the wider literature. The reader should bear in mind the earlier comment that other authors may use slightly different definitions and he should be prepared to translate from one to another in his reading. Key words are shown in italics in the rest of this section.

A *system* is an object of study (or system of interest) which is a collection of components, many of which are related to each other; that is, the components will often be coupled or interact with each other in various ways. Some components may also be coupled to other systems or system components outside the system of interest as defined. The 'rest of the world' outside the system is known as its *environment*. It is sometimes convenient to think of the system and its environment as being separated by a *boundary*, though this need not be the usual kind of spatial one. If system components and environment components are coupled in some way, then the system is coupled to its environment. If part of the environment (containing such components) of the system of interest is defined as another system, then we can see how two systems may be coupled. Such couplings may consist of a flow of components between system and environment, or between systems. In such cases, the flowing components are part of both system and environment, and this shows how systems can 'overlap'.

A system with no flows across its boundary is known as a *closed* system; if there are such flows, it is known as an *open* system. The flows into an open system are known as *inputs* and the flows out are known as *outputs*. Most real systems are, of course, open systems, though it is sometimes a convenient approximation to assume closure. It is also sometimes convenient for the analyst to attempt to define his system of interest in such a way as to make it 'as closed as possible'. Some words of caution are appropriate here to illustrate how different authors use different definitions; some writers, Chorley and Kennedy for example, define a closed system as one with no cross-boundary flows *other than energy flows*. On

21

this definition, the earth is a closed system, while on the earlier one it would be an open system as it receives energy from the sun.

An essential feature of the notion of 'system' and the associated concepts introduced so far is that it is relatively arbitrary. The analyst lists the components to be in his system of interest, and the rest are then in the environment, possibly itself subdivided into other systems. In many cases, 'coherent' or 'whole' objects of study almost define themselves as systems in a natural way. In some cases, there will be an existing literature on how some kinds of system can be defined—for example, that on formal, functional, and nodal regions in geography. In all cases, however, analytical convenience is important.

It is easy to see how the concepts introduced above were used informally in the examples of Chapter 1. In the case of the moorland ecosystem, the system of interest was made up of components consisting of plants and animals in an environment made up of the atmosphere and soil together with energy from the sun. This defines part of the system's boundary; the remaining part would be the spatial boundary of the moorland region being studied, and this would be defined in a variety of ways according to circumstances. The system's components interact through competition for resources, animals eating plants and so on. The coupling to the environment is through the flows in the water and mineral cycles (and through human interference). These flows define the inputs and outputs of the system.

The way in which a system is viewed and defined for study depends on the *level of resolution* (or *scale*) which is adopted. A component which is seen as whole and indivisible at a coarse level of resolution may have an internal structure of its own and be seen to be made up of a number of components, that is itself be a system, at a finer level of resolution. Again, as with system definition, a level of resolution appropriate to the purposes of the analysis in hand should always be chosen. The definition is a matter of analytical convenience—both in relation to the effective formulation of problems posed and to the ease with which a solution can be obtained. As other concepts are defined below, it will be assumed that an appropriate level of resolution has been chosen. It is sometimes useful to distinguish three dimensions of choice of scale: spatial, temporal, and sectoral. The spatial level of resolution refers to the number of spatial units, or zones, to be used in the analysis. Depending on his interest, an ecologist may accept an entire hectare of moorland as his system of interest—a coarse spatial level of resolution—or he may subdivide into a grid of $10\,m^2$ cells—a much finer scale. The temporal scale obviously refers to the time units to be used. These could range, in terms of our examples, from ten-year periods in a dynamic urban systems study which relied on decennial census data, to minutes in a study of the 'recovery' of oxygen levels downstream following the dumping of a pollutant in a river. Sectoral resolution is concerned with the number of component types. At a coarse scale, one may work with the entire population of a city, or the plant biomass of an ecosystem. At a finer scale, the population would be subdivided into types, say according to occupation and income group, or the plant population into species. It is sometimes convenient to define two or more levels of resolution, and it is then usually best if they are related in a *hierarchy*, as with Beer's (1967) cones of spatial resolution shown in Figure 2.1. Note that differences between disciplines within the environmental sciences are often determined by scale of approach. An ecologist studies communities of plants in a particular area; a botanist may study a single plant in detail; a microbiologist may study a single cell in a plant.

The next step in the argument is to define the *state* of a system (at a particular level of resolution). Each characteristic of the system (defined mostly in terms of characteristics of components and couplings but not necessarily entirely) is labelled in some way. A state of the system is then a set of specific values for these labels. This can often be achieved very conveniently by defining an algebraic variable for each characteristic, and then a state is defined by a set of values for

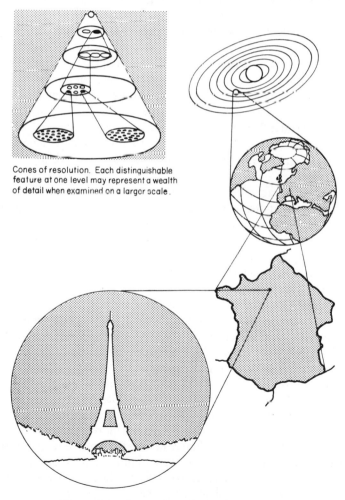

Cones of resolution. Each distinguishable feature at one level may represent a wealth of detail when examined on a larger scale.

Figure 2.1 Cones of resolution (Beer, 1967)

these variables. This, of course, is simply a flexible labelling system of a rather general kind. Henceforth, we will mostly use the term *variable* for a characteristic of a system, even though it may not be algebraic. Let us take a simple example; suppose an index i, which can take the values of 1, 2, 3, ..., N, is a label for species of plant. Let w_i be the biomass of species i in some given system. Then the state of the system is represented as a set of variables $(w_1, w_2, w_3, ..., w_N)$, and a particular state will be represented by such a list with the w_i's taking numerical values. It is also often useful to have a notation for the state of a system at some time t. This can be done with an additional label: $w_i(t)$ becomes the biomass of species i at time t and $(w_1(t), w_2(t), ..., w_N(t))$ the system state at that time. Many more examples will be given in subsequent chapters.

System variables are known as *internal* variables. *External* variables are those which refer to attributes of the environment. The latter may also sometimes be referred to as independent variables. Usually, in state definitions for systems, we will concentrate on defining variables for characteristics which change over time. The system has other 'fixed' characteristics which describe aspects of its *structure* (see below). As the quotation marks indicate, there is an arbitrariness of definition here: structural characteristics may be relatively fixed, but may in reality change slowly over time — the path of a river is an example. This is also an area where other authors occasionally use different definitions, and caution should be exercised. Von Bertalanffy (1968), for example, uses 'external variable' to denote a system property which is measurable and 'internal variable' to denote one which is not. Environmental variables are, however, then defined, as here, as independent variables.

It is intuitively clear that the number of variables needed to describe a particular system, and the number of possible values each variable can take, between them are measures of the size and complexity of the system. The number of variables is sometimes called the number of *degree of freedom* of the system. The overall complexity is represented by the *variety* of the system, defined as the number of possible states which the system can possibly be in. Suppose a system is characterized by a set of variables $a_1, a_2, a_3, ..., a_N$, and the a_ith variable can have n_i possible values. Then the number of possible states is the product $n_1, n_2 ... n_N$. Usually this is a very large number, and, typically, it increases geometrically or exponentially with the number of variables, N. A distinction is sometimes made (for example by Klir and Valach, 1967) between the *full variety* as the number of states the system can achieve in principle (and as defined above) and the *actual variety* as the number of states which can be achieved in practice given that *constraints* will make particular states unattainable.

A convenient distinction can be made between the *structure* of a system and its *behaviour*. The structure is the set of components and the relationships between them; the behaviour is an account of the change in time from one state to another. Note that this definition of structure covers more than *fixed* structure, which arose in the discussion of system variables above. Clearly the system's structure and its environment between them determine the *possible patterns of behaviour*. Sometimes, as we shall see, we will have information on structure and wish to infer something about behaviour; sometimes, vice versa. A particular pattern of behaviour is sometimes called an *activity* of the system.

The notion of behaviour as changes of state can be represented graphically as a path in *phase space* (sometimes called *state space*). Suppose at a very coarse level of resolution (and in this case involving a very crude approximation) we can represent the state of our moorland ecosystem by two variables x_1 and x_2 which are the total plant and animals biomasses respectively. This can be considered to be a prey–predator system which we will consider in more detail in Chapter 8. The behaviour of such systems is usually an oscillation. Phase space is then simply a graphical representation of the state variables in a Cartesian coordinate system as shown in Figure 2.2. A point in this space is a possible state of the system. A curve is an account of system behaviour. The ellipse ABCD, for example, shows oscillatory behaviour. At point A, the animal population (x_2) is a minimum and the plant biomass (x_1) grows with the animal population as the trajectory moves towards B. The animal population has then reached such a size that the plants do not have time to develop properly and the plant biomass starts to decrease. This takes us to C. The plant biomass cannot now sustain the animal population which declines as the trajectory moves to D and further declines to its minimum again at A, though from D to A the plants begin to grow again. We will often find the concepts of phase space and behaviour as trajectories in this space useful. However, most real systems will need more than two variables to describe them, say n, and phase space is then an n-dimensional space which has to be considered in a more abstract way for $n > 3$!

The concept of phase space connects to that of variety introduced earlier. Variety was essentially defined above as a count of the number of points in phase space representing possible states of the system. For continuous variables, a suitable definition of variety will be a volume in phase space. Earlier, we made a distinction between full and actual variety. It is often a matter of speculation as to whether a system can reach *all* the points in some appropriately bounded phase space. If it can, it is said to satisfy an *ergodic hypothesis*.

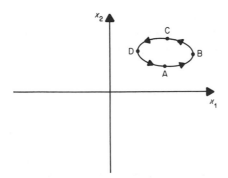

Figure 2.2 State space representation of an oscillating system

The behaviour of a system will be sensitively related to the couplings between components, or sets of components, which can be considered as *stocks*. These couplings will often be in the form of *channels* carrying various kinds of *flows* from and to stocks: energy, materials, money, people, information, or whatever. Much of environmental science is concerned with the *location* of stocks and products and the *flows* and *interactions* between these locations along *routes* (or *pathways*) in channels which are parts of *networks*. So at this point we can anticipate a fruitful marriage between systems analysis and the spatial or geographical aspects of environmental science.

The most important of the couplings for overall system behaviour are various kinds of *feedbacks*. Consider a flow (1) from system A to system B, and another (2) from system B to system A, as shown in Figure 2.3. If flow (2) always follows flow (1), then this is a feedback from system A to itself (via system B). System A could be a reservoir and B a control system. Let flow (1) be a reported measurement of the level of the reservoir and then (2) may operate a valve to refill the reservoir if it is losing water, or vice versa. This is an example of *negative feedback* — the second part of the loop has the opposite effect to the first. In this case, it is being used to maintain the level of a reservoir more or less constant. This is a characteristic of negative feedback loops: they form part of *control systems* to maintain stability. More examples will be presented below in the section on system diagrams.

Suppose now that we have a city divided into two regions, the inner city and the suburbs. Let system A be the inner city population and system B the suburban population; flows (1) and (2) are considered to be information flows between the two systems. Suppose further that the quality of life in the inner city deteriorates if people leave (through loss of local tax income) while the suburbs can expand indefinitely. Then if (1) contains reports of a poor quality of life, and (2) of a good quality of life, people will tend of move from A to B; but the quality of life in A will deteriorate yet further and so yet more movements out will occur. This is a *positive feedback* loop, where the phenomenon being affected by the loop is reinforced. This sort of behaviour is usually ultimately unstable.

This is a convenient point to note that a component, or particular groups of components, within a system may have coherence, and exhibit 'collective' behaviour and be analytically convenient: that is, they behave like, and are, systems. In this case they can be called *subsystems* of the original system. They themselves may have subsystems: indeed a whole hierarchy of interlocking

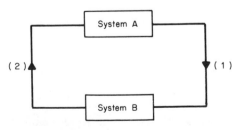

Figure 2.3 Feedback

(a) Stable   (b) Unstable   (c) Neutral

Figure 2.4   Types of equilibrium

systems may be defined, in effect a hierarchy made up of different levels of resolution. It is often useful, when some system of interest is being defined, to seek to identify a useful subsystem structure at the outset. These decisions can often be recorded in the sort of diagrams to be discussed in the next section. Such subsystems are likely to refer to different types of component — for example population groups and economic sectors within a city. But they may also be useful if the system has a complicated set of feedback loops. These may occur within subsystems, between subsystems, or between the system and its environment. Even more complicated structures can be discovered or defined for a more extended hierarchy of subsystems.

We have now introduced a number of concepts which enable us to describe system structure and behaviour. The final step in this initial presentation of basic concepts is to introduce some of the notions associated with different types of system behaviour. These are, usually, idealized notions when compared with real-world systems, but nonetheless offer considerable insight and clues to the analyst. A system may persist in one state for quite a long time, and this is likely to be an *equilibrium* state. There are three kinds of equilibrium exhibited for a simple mechanical example in Figure 2.4: stable, unstable, or neutral. Figure 2.4(a) shows a ball in an inverted hemispherical bowl. It rests in the bottom, which is a position of stable equilibrium. The essence of this is that if the ball is displaced a short way, it returns to its original position. The other two figures are now almost self-explanatory: in Figure 2.4(b), a small displacement of the ball leads to it travelling a considerable distance at an increasing velocity; in Figure 2.4(c), a displaced ball stops at the new location. In the stable case, gravitational forces and pressures from the bowl exert negative feedback following a displacement to maintain equilibrium; in the unstable case, they exerts positive feedback; in the neutral case the forces have no horizontal component. This is typical: stable equilibrium is usually associated with some kind of negative feedback, and vice versa. Systems in stable equilibrium states maintained by negative feedback forces are said to be *homeostatic*.

This kind of system behaviour can be exhibited on phase space diagram. Let x_1, x_2 be coordinates in a plane at right angles to the plane of Figure 2.4. This is shown as Figure 2.5. Then the position of the ball *projected* onto this plane defines the state of the system and the (x_1, x_2) space constitutes a phase space Typical trajectories for the three situations are shown in Figure 2.6. In case (a), if the ball is in a disequilibrium position, it is attracted to the stable point (which is hence known as an *attractor*). Conversely, the equilibrium point in (b) is a *repeller*. In (c), the trajectories can be straight lines in any direction.

28

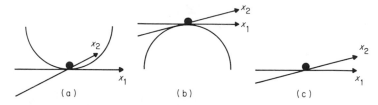

Figure 2.5 System coordinates for Figure 2.4

Experience tells us that these concepts of equilibrium can be applied to parts or wholes of more complicated systems. The total population in a small 'stable' (say suburban) area of a town is likely to be stable and in equilibrium for quite a long time. A moorland ecosystem, however, as we have seen, is not likely to be in equilibrium. It will either oscillate or progress through succession to a new kind of state altogether. The river system part of our water resource example can be used to illustrate another kind of equilibrium—a type of dynamic equilibrium known as a *steady state*. In this case, system components are in motion, as water down the river, but the flow can often be considered to be more or less constant (at least in long run averages). This is an example of steady state equilibrium, maintained in this case by the hydrological cycle. A different type of steady state involves the maintenance of a stock or level, for example the water level in a reservoir, by balanced inputs and outputs.

We have seen that all kinds of stable equilibria are maintained by some 'forces' of negative feedback. This often involves some kind of underpinning *optimizing principle*: people maximizing utility, water in a river taking the least work path, and so on. The ball in Figure 2.4 seeks the position of minimum energy. The forces which maintain equilibrium can be seen as generated by such optimizing principles.

Typically, systems will not be in equilibrium. If they have equilibrium states, and are tending towards these following some disturbance, then the intermediate states are called *transient states* and the average time taken to reach equilibrium is known as the *relaxation time*. More complex and highly organized systems are likely to have larger relaxation times. In many cases, however, the system will not have equilibrium states and the trajectories in phase space will be much more complicated. This issue will be taken up again in Chapter 8.

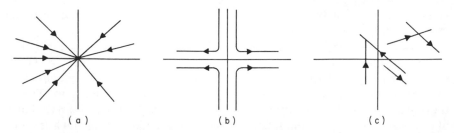

Figure 2.6 Trajectories in state space for different types of equilibrium: (a) stable; (b) unstable; (c) neutral

It is appropriate to conclude this enumeration of basic concepts with that of a *model* of a system. The analyst attempts to represent his knowledge and understanding about a system in a theory, as we have seen. A model is a formal representation of such a theory. The formality will usually be introduced through the language of mathematics or computer simulation. Related concepts are *isomorphism* and *homeomorphism*. Two systems are said to be isomorphic if there is a one–one relationship between them—that is between their components, or, more generally, their structures. The relationship is homeomorphic if it is many–one or one–many rather than one–one; that is, broadly speaking, the systems are related but not precisely. The standard topological map of the London underground system, for example, bears a homeomorphic relationship to the real network. The goal of the analyst could then be said to be to build a model which is isomorphic to his system of interest, though in practice it will be homeomorphic. The nearer it can approach isomorphism, the better.

2.2 SYSTEM DIAGRAMS

It is often convenient to attempt to represent the main features of the structure of a system in a diagram. Many authors do this, but often using different conventions. The procedure in this section, therefore, is to attempt to identify the main features which are incorporated in diagrams, to show the symbols different authors use for these, and to provide the reader with enough information to be able to construct diagrams to his own taste for his own systems of interest.

The main features represent different types of system component and can be summarized as follows:

(1) *Stock/storage/level/property* For the ecosystem example, such variables would include biomass; within a water resource system, the levels of water in reservoirs; and in a city, populations of different types, jobs, housing stocks, and so on. Such variables may themselves be major system components, like a volume of water. Quite often, as in some of these illustrations, such stock variables are 'counts' of system components—such as people of a particular type. Some stock variables are sources or sinks within the environment, and these are sometimes represented by distinctive symbols, such as Forrester's 'clouds'.

(2) *Related structural features* A number of fixed parameters often need of be represented as structural features. This may be the average hours of sunshine received by an ecosystem, the capacity of a reservoir, the cost of building houses, and so on.

(3) *Coupling/flow/interaction* Flows are usually to and from stocks of some kind and are carried in channels. A number of examples were given in the previous section. They include information flows as well as material or person flows.

(4) *Regulator/valve/function* These are system components which control a flow

30

in or out of a part of the system. A natural regulator in the ecosystem is the degree of heath 'cover' which controls the amount of light reaching plants at ground level. The flow of water out of a reservoir is controlled by a valve. The city has 'softer' regulators, such as planning controls or traffic management schemes.

(5) *Subsystems* It is often convenient to exhibit subsystems explicitly in diagrams. Again, examples have been given earlier. It may be useful to identify different types of subsystem with different functions, as we will see below.

(6) *Inputs and outputs* These are particular flows into and out of components or subsystems, or indeed the system itself (to connect it to its environment). Many authors thus use special symbols for these flows, as we shall see.

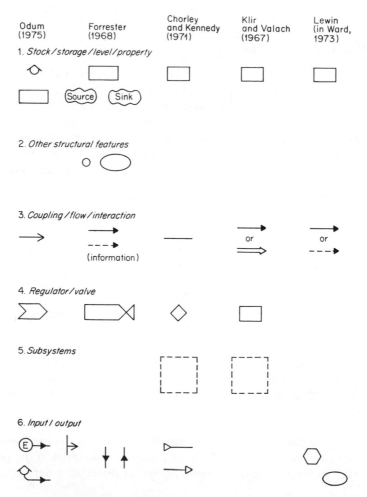

Figure 2.7 The main symbols for system diagram

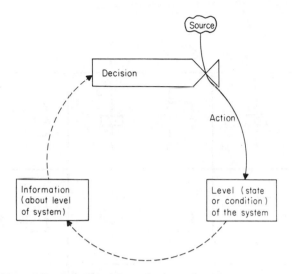

Figure 2.8 A feedback loop in the style of Forrester (1968)

The special symbols used for these various elements of systems by five authors (or pairs of authors) are presented in Figure 2.7. Some, such as Odum (1975), have a different symbol for most features; others, such as Klir and Valach (1967), use mainly squares and arrows and give their diagrams a more abstract appearance. Some examples of the work of these various authors is given below. Figure 2.8 shows a feedback loop in the style of Forrester (1968). Figure 2.9 shows some

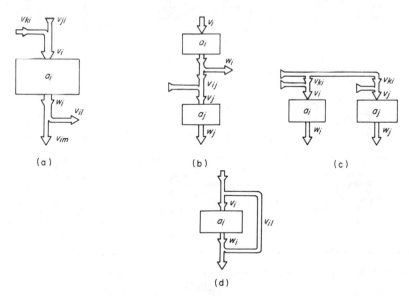

Figure 2.9 Couplings, in the style of Klir and Valach (1967). (a) Method of indicating couplings between elements; (b) series coupling; (c) parallel coupling; (d) feedback coupling

32

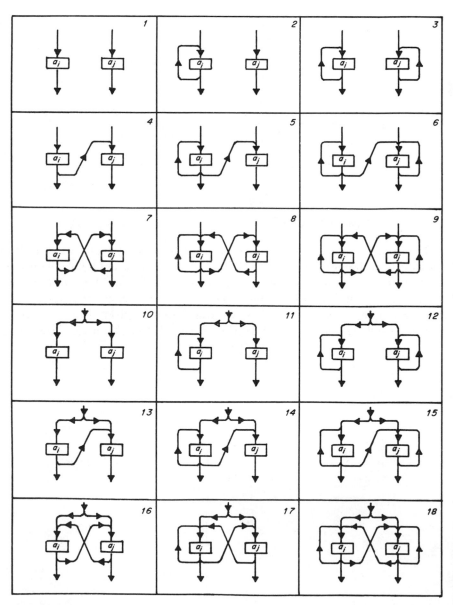

Figure 2.10 Types of coupling between two elements of a system (Klir and Valach, 1967)

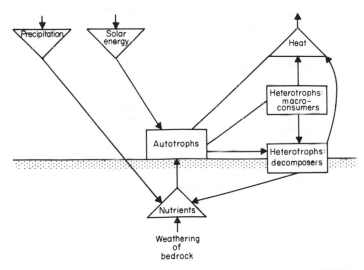

Figure 2.11   An ecosystem, after Chorley and Kennedy (1971)

simple couplings in the style of Klir and Valach. Figure 2.10 shows the variety of ways of coupling two elements in the same style.

We can then go on to examine examples of system diagrams which are closely connected to our three original examples. One has already been given, as Figure 1.11, but it is repeated here for convenience: that is an ecosystem in the style of Chorley and Kennedy in Figure 2.11. This can be compared with Odum's diagram, which can also be interpreted as an ecosystem model, in Figure 2.12, with P_1 as the plants, P_2 as herbivores, and P_3 as herb/carnivores. The water example can best be illustrated by two models of the hydrological cycle: the first by Odum, which shows how to convert from a picture model to something more formal (Figure 2.13) and the second by Lewin, presented by Ward (1973) (Figure 2.14).

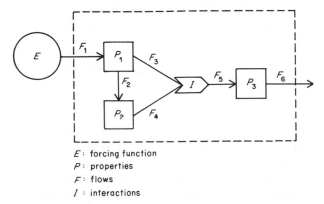

E : forcing function
P : properties
F : flows
I : interactions

Figure 2.12   An ecosystem, after Odum (1975)

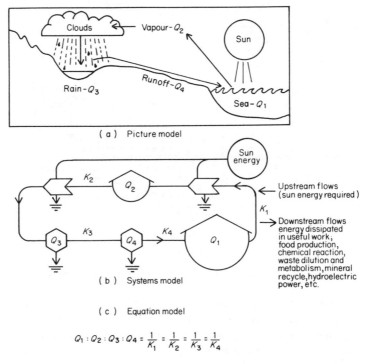

(a) Picture model

(b) Systems model

(c) Equation model

$$Q_1 : Q_2 : Q_3 : Q_4 = \frac{1}{K_1} = \frac{1}{K_2} = \frac{1}{K_3} = \frac{1}{K_4}$$

Figure 2.13 The hydrological cycle, after H.T. Odum (E.P. Odum, 1975)

In the urban case, Figures 2.15 and 2.16 show aspects of Forrester's *Urban dynamics*. Figure 2.15 shows the nine major stock variables—three categories each of economic activity, housing, and labour force—and Figure 2.16 shows how the influences controlling a rate variable are exhibited. This means that the whole system diagram is immensely complicated, of course.

Although considerable insight can be gained from system diagrams, it is appropriate to conclude this section with a short discussion on their limitations, especially in relation to sciences involving spatial analysis. But first, a comment which is generally applicable: the diagrams developed so far generally neglect time. They are mostly concerned with structure, and hardly at all with behaviour. It should be conceded, however, that progress with this problem will be difficult, as is indicated by the limitations of phase space diagrams and time trajectories to very simple systems involving two or three dimensions only.

In order to incorporate space explicitly, a marriage is needed between traditional mapping and the kinds of diagrams presented earlier. This is more easily solvable in principle than the time problem, but if system components are plotted diagrammatically on a map base, because of the coincidence of activities at particular locations, 'too much' would have to be plotted for comfort. Nonetheless, some progress could be anticipated in such a direction. It may also

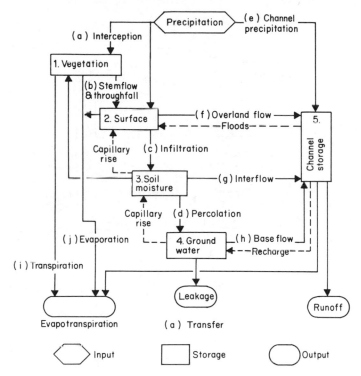

Figure 2.14 The hydrological cycle, after Lewin (Ward, 1973)

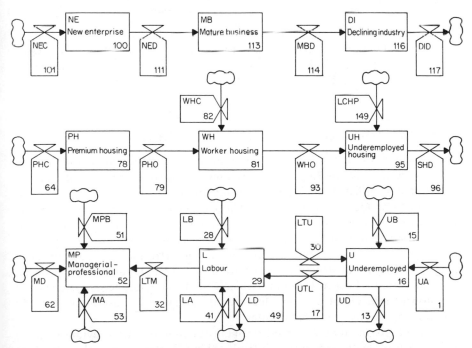

Figure 2.15 The main stock variables in Forrester's urban system (Forrestes, 1969)

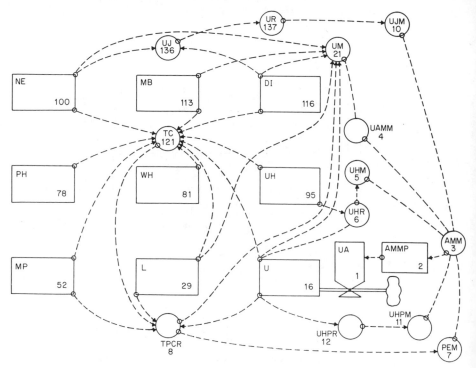

Figure 2.16 The influences controlling a rate variable in part of Forrester's urban system (Forrester, 1969)

be useful to integrate system diagrams with the space–time diagrams of the Lund school of geographers in Sweden. They use the three available dimensions to represent time and the two spatial dimensions of a map, and plot lifelines of system components in such diagrams.

2.3 CONCEPTS AND DIAGRAMS: A SUMMARY

It is useful both to summarize the concepts introduced in Section 2.1 and to relate them to the elements of the system diagrams presented in the previous section. The concepts are listed in Figure 2.17 in the order in which they are presented; this should provide a useful check list. Those which correspond wholly or partly to the six types of element for system diagrams are marked with an asterisk and the corresponding number. The correspondences are exact for (1) stocks, (3) couplings, (5) subsystems, and (6) input and output. Parameters (2) are, in effect, exogenous features of structure and so are shown there; regulators or valves (4) are special kinds of control systems (which would, of course, be *sub*systems) and are shown there. Other concepts featured in the summary are implicit in the diagrams or the theories which underpin them.

The concepts and diagrams have been illustrated in the two previous sections with aspects of the three examples introduced in Chapter 1. These three kinds of

system

environment

boundary

closed/open

inputs/output*6

level of resolution

state

variables (internal, external, independent)

degrees of freedom

variety (full, actual)

(structure → including parameters)*2

behaviour

activity

phase (state) space

couplings*3

channels

stocks*1

flows

routes (pathways)

networks

feedback

(including → control systems regulators)*4

subsystems*5

equilibrium (stable, unstable, neutral)

homeostasis

steady state

optimizing principle

transient state

relaxation time

model

isomorphism

homomorphism

Figure 2.17 Basic concepts

system are each so complex that it is impossible to go further than this. The concepts will all be used, implicitly or explicitly, in various ways in subsequent chapters on methods. The main purpose of the diagrams is to seek initial insight, and to summarize elements of theories about particular systems. Other methods will usually be needed to represent systems in more detail; these are the subject of Part 2.

2.4 TYPES OF PROBLEM IN SYSTEMS ANALYSIS

Klir and Valach (1967) usefully identified three kinds of problem as the concerns of systems theory. First, there is *analysis*, which is concerned with the investigation of system structure and the prediction of behaviour. Secondly, *synthesis* is concerned with the design of a system or subsystem structure which will achieve some desired behaviour. This involves the synthesis of a suitable set of components. Thirdly, there are '*black box investigations*' which arise when the structure of the system is unknown but much information is available on system behaviour. Particular attention is then paid to the inputs and outputs of the system and (assuming that analysis of structure is infeasible) attempts made to predict the latter given the pattern of the former. Sometimes it may then be possible to make inferences about the structure of the black box (which is the unknown system structure).

Analysis is more or less identical to the normal activities of the scientist — except that, as we noted in Chapter 1, the system theorist will have a particular concern with large complex systems whose elements are highly interdependent. The problems of synthesis have also been well known in engineering and other applied science and planning fields. This activity is a more formal expression of the aim noted in Chapter 1 that the products of systems theory would be useful in planning. The relatively new feature of this work again lies in size and complexity. Systems engineers have learned how to assemble complex control systems which can accomplish a large range of tasks from relatively simple elements and the methods for doing this have been developed within a 'systems' approach. The 'black box' problem may at first be thought unique to systems theory, but it too has parallels in other branches of science — for example with 'behaviourism' in psychology. We return to the question of 'synthesis' and its connection to planning in Chapter 9.

We will continue to use the term 'systems theory' to cover work on all of these problems, but in a particular situation the reader should always consider carefully what type of problem, or combination of types of problem, he or she has.

In Chapter 1, we distinguished between inductive and deductive methods in science and inclined towards favouring the second of these approaches for the advanced sciences. There is a sense in which systems theory, in its search for generality and the transferability of *methods* from one particular system to another, does, in an idealized way, represent an inductive approach to theory building. The traditional inductive method in science is to attempt to infer general theories from data — from direct observations. Systems theory may be seen as an attempt to offer appropriate methods for theory building, and hence virtually a metatheory, from observations of the *type of system* being studied. This is an ideal, and is far from being achieved, but it does add a 'higher level' kind of problem to systems theory and sets the scene for a discussion of types of system in the next section.

2.5 TYPES OF SYSTEM

To seek increased generality involves looking at particular systems in more abstract ways. Systems which are very different in many particular features may then appear to be 'similar' in certain abstract ways. There is a spectrum of levels of abstraction, ranging from a concern with the particular and unique to the completely general and abstract. Nearly always, we operate at various intermediate levels.

It is in this spirit, of seeking some generality by examining certain abstract

Figure 2.18 Weaver's classification of systems: (a) simple; (b) disorganized complexity; (c) organized complexity

characteristics of systems, that we seek to classify *types*. There is no unique classification of course. Different authors have classified systems in various ways for several purposes. The different schemes relate sometimes to the complexity of the system and sometimes to aspects of function. We discuss in turn the classification introduced by Weaver (1958), Chorley and Kennedy (1971), and Klir and Valach (1967).

Weaver illustrates his analogy with three different examples of billiard balls on a table, as shown in Figure 2.18. In the first case there is a single ball on the table, in the second a large number, and in the third a large number but with the additional complication that there are many elastic-band connections between the balls.

The first type of system is called *simple*. It is characterized by a very small number of components and can be described by a correspondingly small number of variables: in the case of the single billiard ball, four, for position and velocity in a plane. Weaver argues that science had been largely dominated by this kind of problem until late in the nineteenth century. If the billiard ball is disturbed, for example, its motion can be predicted using the methods of Newtonian mechanics.

The second type of system is of *disorganized complexity*, characterized by a large number of components and variables, but with couplings between them which are weak or random. At first sight, it would seem that if the corresponding billiard ball system was disturbed, the motion of its components could be predicted by Newtonian mechanics, but this turns out not to be the case because of the complexity: there are simply too many mathematical equations to handle even with modern computers. This is even more obviously true for other systems of this type. The number of particles in a standard volume of a gas is the order of 10^{23}; the population of a large city may be greater than 10^6. In the late nineteenth century, scientists such as Boltzmann and Gibbs showed how certain quantities of interest relating to such systems could be calculated using a new style of mathematics—essentially a statistical averaging, or more technically, an entropy maximizing process. Although it was impossible to predict the behaviour of any one particle in a gas, the distribution of particles among energy states could be calculated, as could more aggregate quantities such as the pressure and temperature of the gas. Indeed, it turned out that the more aggregative laws of thermodynamics could be predicted in this way. The equivalent comment for the billiard balls example is that, while the behaviour of one ball could not be predicted, the distribution of velocities, could, together with certain known quantities such as the average rate at which a ball strikes a cushion. This method of analysis, however, remained very much in the realms of physics until the 1940's when Shannon and Weaver's (1945) work on information theory began to suggest wider applications. These methods are outlined in more detail in Chapter 4.

There is also a more dynamic type of analysis which can be applied to systems of disorganized complexity — a different kind of statistical averaging procedure. If we are interested in the rate of change of, say, the number of system components moving from a given state to another, such as the number of people dying in some time period, then it seems reasonable to compute and to work with the average rates of change and to build models based on these. These methods are based on principles of accounting and are described in Chapter 6. Among the first methods of this type were those developed by Markov and it is interesting to consider that he was publishing his results at about the same time as Gibbs, in the early twentieth century. Perhaps that time was ripe for the development of statistical averaging methods on both fronts.

The third type of system in Weaver's classification is of *organized complexity*. These systems are large, as with the second type, but now have the additional complication of some strong couplings between the elements — as represented by the elastic bands in the billiards example. This provides a high degree of organization which is not present in type-II systems — and hence their respective names. Weaver's argument, which dates back at least to the 1940's, was that many of the severe problems of modern science are of this type; the argument is probably still true today. A good example is provided by the brain and its neurons, mentioned earlier: the 10^{10} neurons form a highly organized whole, and while much may be known about micro-structure, relatively little is known about this whole. Some methods have been developed for tackling problems of this kind, and they will be described later, mainly in Chapters 8 and 9. However, it is likely that there are many types of problem within this class and relatively little general progress has been made.

Since we have argued that systems theory is essentially concerned with complexity, Weaver's types II and III are of the most important. It will already be clear that considerable insight can be gained, together with some clues about which methods of analysis are likely to be most applicable, if the Weaver type of a system can be established; this is a procedure which will often be used subsequently.

Chorley and Kennedy, arguing in the context of physical geography, identify four system types. They are concerned not so much with the complexity of the system, but with the way in which the analyst chooses to look at it (though they do note 'a progressive sequence to higher levels of integration and sophistication'). The four types of system are illustrated for simple cases in Figure 2.19. First, they define *morphological* systems. As the name implies, this means that the analyst is largely concerned with structure, and the technique of analysis they suggest is the identification of structural relationships using statistical correlation analysis. Their second type is the *cascade* system, for which the analyst is concerned with flows, such as energy or matter, through the system, with the identification of pathways which carry such flows, and with storage or

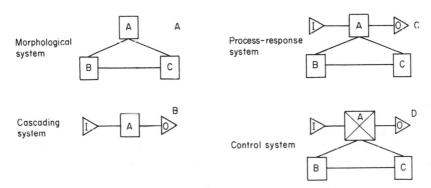

Figure 2.19 Chorley and Kennedy's system types

interaction units in such networks. Thirdly, there is the *process-response* system which corresponds closely to the concept of 'black box' introduced in the previous section. Finally, they identify *control* systems and a concern with the structure of feedback loops which generate particular control functions, and possibly with the synthesis of such systems for particular management purposes.

Klir and Valach identify types associated with system function. They begin with *communication* systems, which are the structures that transmit signals across space. These are a version of Chorley and Kennedy's cascade system. Secondly, they introduce *memory* systems, which are capable of storing information. Thirdly, they note that some systems are *decoders* which generate a one–one relationship between a set of inputs and corresponding outputs. Fourthly, they introduce the notion of *function generators* which transform a set of inputs to produce outputs which are some functions of these. Such systems are equivalent to the 'regulators' or 'valves' of Section 2.2. The inverse of these are *retrospective* systems: where unique inputs may be associated with any output. Finally, they identify, as do many other authors, *self-organizing* systems. These are special kinds of control system whose behaviour can be adapted from within the system itself. This is obviously closely connected to the notion of systems which can 'learn'. We will see examples of these in Chapter 9.

A more elaborate classification of systems (or subsystems, as appropriate) by function is offered by Miller (1978). His argument is that there are nineteen types of critical subsystems of any living system. The first two of these have rather broad functions: the reproducer, the subsystem which provides the mechanism for the system to produce others like itself; and the boundary, the perimeter of the system which includes such functions as protection from stress and the regulation of the flow of matter, energy, and information to or from the system. Since these subsystems will usually be made up of sets of more basic subsystems, which form the other seventeen types, we focus for the time being on these. They are listed in Table 2.1 and we have retained Miller's numbering.

Miller makes a fundamental distinction between subsystems which process matter and/or energy on the one hand and those which process information on the other. There is an obvious common-sense distinction here: for example, an

42

Table 2.1 The 19 critical subsystems of a living system

SUBSYSTEMS WHICH PROCESS POIN MATTER-ENERGY AND INFORMATION

1. *Reproducer*, the subsystem which is capable of giving rise to other systems similar to the one it is in

2. *Boundary*, the subsystem at the perimeter of a system that holds together the components which make up the system, protects them from environmental stresses, and excludes or permits entry to various sorts of matter-energy and information.

SUBSYSTEMS WHICH PROCESS MATTER-ENERGY	SUBSYSTEMS WHICH PROCESS INFORMATION
3. *Ingestor*, the subsystem which brings matter-energy across the system boundary from the environment.	11. *Input transducer*, the sensory subsystem which brings markers bearing information into the system, changing them to other matter-energy forms suitable for transmission within it.
	12. *Internal transducer*, the sensory subsystem which receives, from subsystems or components within the system, markers bearing information about significant alterations in those subsystems or components, changing them to other matter-energy forms of a sort which can be transmitted within it.
4. *Distributor*, the subsystem which carries inputs from outside the system or outputs from its subsystems around the system to each component.	13. *Channel and net*, the subsystem composed of a single route in physical space, or multiple interconnected routes, by which markers bearing information are transmitted to all parts of the system.
5. *Converter*, the subsystem which changes certain inputs to the system into forms more useful for the special processes of that particular system.	14. *Decoder*, the subsystem which alters the code of information input to it through the input transducer or internal transducer into a "private" code that can be used internally by the system.
6. *Producer*, the subsystem which forms stable associations that endure for significant periods among matter-energy inputs to the system or outputs from its converter, the materials synthesized being for growth, damage repair, or replacement of components of the system, or for providing energy for moving or constituting the system's outputs of products or information markers to its suprasystem.	15. *Associator*, the subsystem which carried out the first stage of the learning process, forming enduring associations among items of information in the system.
7. *Matter-energy storage*, the subsystem which retains in the system, for different periods of time, deposits of various sorts of matter-energy.	16. *Memory*, the subsystem which carries out the second stage of the learning process, storing various sorts of information in the system for different periods of time.

17. *Decider*, the executive subsystems which receives information inputs from all other subsystems and transmits to them information outputs that control the entire system.

18. *Encoder*, the subsystem which alters the code of information inputs to it from other information processing subsystems, from a "private" code used internally by the system into a "public" code which can be interpreted by other systems in its environment.

8. *Extruder*, the subsystem which transmits matter-energy out of the system in the forms of products or wastes

9. *Motor*, the subsystem which moves the system or parts of it in relation to part or all of its environment or moves components of its environment in relation to each other.

19. *Output transducer*, the subsystem which puts out markers bearing information from the system, changing markers within the system into other matter-energy forms which can be transmitted over channels in the system's environment.

10. *Supporter*, the subsystem which maintains the proper spatial relationships among components of the system, so that they can interact without weighting each other down or crowing each other.

---

individual eating is absorbing food and energy; the same individual reading a book is absorbing information. However, the processes through which information is absorbed and acted upon each involve matter and energy, and so, in a sense, the information subsystems are special cases of the matter–energy subsystems. However, the basic distinction is nearly always clear. The only blurred area is likely to be at the biological end of our spectrum when it is a matter of convenience as to whether, say, plant development is seen wholly in matter–energy terms or partly to include information processing—for example, the 'information' in plant DNA which contains the programme for growth.

It is also important to note Miller's argument that these functions can be identified from all different scales from the microbiological through to all kinds of organizations—even nations. For our present purposes it forms a useful checklist: we can identify different types of system components in relation to it and see whether anything has been missed.

Miller's nineteen system types, and certainly the seventeen basic ones, can easily be related to the six kinds of system components discussed earlier and listed in Figure 2.7. The *ingester* and *extruder* are input and output devices, the *distributor* is a coupling *matter-energy storage* is a stock. The remaining Miller types (on the matter–energy side of the diagram)—the *converter, producer, motor,* and *supporter*—focus on particular subsystem functions which involve labelling and distinguishing these types among the 'subsystems' of the diagrams generated by the methods of Figure 2.7. Similar analyses can be offered for Miller's information-processing system types.

NOTES ON FURTHER READING

In Chapter 2, we attempted a synthesis of ideas culled from many sources. There is no single general book on systems theory which seems to offer an adequate background for the perspective of this book. *Forrester (1968, 1969), *Klir and Valach (1967), and *Miller (1978) all offer useful ideas. The next stage is to seek other geographical texts which offer support. †Chorley and Kennedy (1971) is an interesting book based on a specific classification of systems. †Huggett (1980) offers an elementary introduction which will be useful supplementary reading. †Chapman (1977) provides a careful scrutiny of many of the concepts which are introduced more loosely in this book and is particularly recommended as complementary reading. *Bennett and Chorley (1978) use the more advanced concepts of systems analysis as applied mainly in fields like engineering and apply them to environmental systems. †Jeffers (1978) reviews systems analytical methods from an ecological viewpoint.

It is also useful to mention some of the classical papers used by geographers in defining systems and related techniques (†Hall and Fagen, 1956; †Ackerman, 1958, 1963) and which have adopted a more or less equivalent ecological perspective (†Barrows, 1923, †Stoddart, 1965). The application of general systems theory in geography is discussed with some scepticism by †Chisholm (1967).

A specific back-up reference for Section 2.5 on system types is †Weaver (1958, pp. 7–122).

PART 2

Methods of System Theory and Spatial Analysis

CHAPTER 3

A 'methods' perspective for systems analysis

3.1 SYSTEMS, METHODS, AND SPATIAL ANALYSIS

Most disciplines are involved in studies of particular systems of interest, though in each case they evolve specific methods of study. Disciplines such as mathematics and, perhaps, philosophy are almost wholly concerned with methods. Since, in systems theory, we are attempting to achieve generality, but we recognize the distinctive nature of many systems of interest, the best way to make progress is to adopt a methods perspective — to try to identify the methods which are appropriate to the study of particular types of system (and for this we draw heavily on the last section of Chapter 2).

In Part 2, we identify a number of topics which cover many of the general approaches to system modelling currently available. They progress along a path of increasing complexity, though the order of the next three chapters is relatively arbitrary in this respect. The topics are chosen to reflect certain known characteristics of system behaviour. It should be emphasized, however, that a number of well-known model building methods are available which are useful in systems theory which do not have such connotations. These are mentioned in Section 3.3 and include elementary mathematical modelling, computer simulation, and statistical techniques. There are many textbooks available, of course, on such techniques. The objective in this book is to concentrate on more specific techniques which are associated with some prior knowledge of the type of system being modelled.

Some of the topics arise out of Weaver's classification of systems. We saw in Chapter 2 that systems of disorganized complexity can be modelled by statistical averaging methods — based on entropy maximizing and accounting. These methods are presented in Chapters 4 and 5 respectively. We also saw in the discussion of equilibrium that homeostatic forces are often generated as a result of optimizing behaviour within systems, and that in any case systems often contain subsystems or components which exhibit such behaviour and so Chapter 6 is devoted to optimization. Chapters 4 and 6 both involve an emphasis on the role of *constraints* in describing system structure within a model, while the accounting methods of Chapter 5 are concerned with special forms of constraints, the accounting equations, which are the *conservation laws* of the system. All three approaches are used for system which usually exhibit relatively elementary *dynamic* behaviour, though the models generated often form good approximations for more complicated behaviour. Further, when external variables are allowed to vary, connections can be established, within Chapter 4, to the *second law of thermodynamics* which in a general form may be seen as one of the general theorems of systems analysis — concerned with directions of change and irreversibility for certain kinds of system. (The *first law* can be seen as a special case of a conservation equation.) It is also possible to investigate the *stability* of these elementary kinds of dynamic behaviour.

Network theory is picked out for Chapter 7 because it involves a different kind of procedure than any other branch of systems theory — for example in the task of finding the shortest path through a network, which is necessary as a component of many kinds of system models. It sometimes also provides a useful tool for examining system structure in a general way, as with Atkin's *q*-analysis.

More complicated dynamic systems behaviour is treated in Chapters 8 and 9. First, various aspects of dynamical systems theory are reviewed. Various kinds of Lotka–Volterra prey–predator systems from ecology turn out to be much more widely applicable within the environmental sciences and it is possible to use some recently developed results in the theory of differential equations to gain insight in this respect. Insights are also gained from the recently developed *catastrophe theory*. In Chapter 9 various approaches to control and planning are reviewed, including a discussion of Ashby's *law of requisite variety*, which is a powerful general systems theorem, and Beer's conjectures on the structure of control systems based on neuro-physiological analogies.

3.2 MATHEMATICAL PREREQUISITES

A consistent and serious attempt will be made in Chapters 4 to 9 to keep the mathematical requirements of the reader at a modest level. It is essential, however, that the reader should be familiar with elementary algebra, including ways of defining variables to describe environmental systems (which may, for example, involve many subscripts and superscripts), functions, and graphs and elementary coordinate geometry. This material is covered in the first three chapters of *Mathematics for geographers and planners* (MGP) by Wilson and Kirkby (1975). On the rare occasions when more advanced concepts are needed, they will be defined as appropriate. If it is not possible to do this rigorously or comprehensively, an attempt will be made to convey the basic idea and to give a reference elsewhere for the details.

It is hoped that the reader who does not wish to develop his mathematical skills beyond this minimum will find the book a useful introduction to systems theory. However, the further reading in various chapters is intended to lead into greater depth. For such reading, and for deeper understanding of the material as presented, further mathematical skills will be necessary. For example, a knowledge of elementary matrix methods (Chapter 4 of MGP) will help in the study of accounting methods; a knowledge of calculus (Chapters 5 and 9 of MGP) will be generally helpful, but particularly so for Chapters 5, 6, and 8; the study of differential equations and related methods (Chapters 6 and 10 of MGP) will aid dynamical systems theory.

3.3 A NOTE ON SOME STANDARD METHODS

As noted earlier, there are a number of standard methods available to the systems analyst and these should not, of course, be neglected. The main ones can be summarized as elementary mathematics (particularly algebra and calculus), computer simulation, and statistical (including many so-called econometric) methods. There are large numbers of standard texts available which describe these methods. Broadly speaking, the simpler the system, the more likely it will be that a model can be built using elementary methods. A block diagram could be constructed using the methods of Section 2.2, and the connections in such diagrams will imply functional relationships between the main variables. If these are known in a reasonably precise form, then mathematical analysis will be appropriate. If they are conceptually simple relationships, but numerous or involving a lot of calculation, then computer simulation is available. If the relationships are 'suspected' rather than precisely known, then an appropriate statistical method can be used for elucidation.

For more complicated systems, an initial study may suggest that more specific insights into the model building task can be provided by one or more of the methods described in the following chapters.

3.4 TOWARDS A MATHEMATICAL REPRESENTATION FOR THE SPATIAL ANALYSIS OF ENVIRONMENTAL SYSTEMS

It was established in an informal manner in Part 1 that environmental systems are complicated, and we have argued in this chapter that the methods to be used to handle this complexity are mathematical. At the deepest levels of analysis, the methods involve the most advanced mathematics but, though glimpses of this will occasionally be given, these will be beyond the scope of this book. It is important to realize, however, that much progress can be made using only elementary mathematics *and that the key to this is the way in which systems are represented in mathematical form*. A preliminary introduction to this topic is therefore provided in this subsection.

The most important initial step in any system modelling exercise is the definition of *state variables* (see Section 2.1). A system is made up of its components, couplings, and so on, and these must be described by algebraic variables. The obvious and most detailed way to tackle this problem would be to characterize each constituent part and then to chart the changes in these characteristics over time. The characteristics would usually include location and then a number of indicators of type—say age, occupation, income, house, job, etc., for a person. Even at this level, interdependence between components is evident, because homes and jobs, for example, are themselves system components. Thus, for people, such a representation would consist of a list of characteristics for each person in turn: person $1 - x_{11}, x_{12}, x_{13}, x_{14}, \ldots, x_{1,50}$; person $2 - x_{21}, x_{22}, \ldots, x_{2,50}$ (assuming 50 characteristics). If the system contained 100 000 people, this description would involve 5 000 000 algebraic variables. Note that this list could be written in a shorthand way as $\{X_{p,q}\}$, $p = 1, 2, \ldots, M, q = 1, 2, \ldots, N$ for M people in all and N characteristics in all. This may be called the 'individual-component' representation of systems. It is ideal in that it contains the maximum possible information on the system state, but often impractical in that it involves thousands or millions of algebraic variables. It is occasionally useful in spite of the difficulties (see Wilson and Pownall, 1976; the point will be taken up later in Section 8.5).

The most effective way to reduce the number of state variables is to group components according to various characteristics and to count the number of components in each group. For example, if age, occupation and income are characteristics p, q, r, then K_{pqr} could be defined as the number of people age p, in occupation q and with income r. Let i and j be characteristics which describe residential and workplace location (as zones on a map), and T^q_{ij} be the number of people in occupation q who live in zone i and work in zone j. We can continue to define other state variables in this way. This might be called a 'group-component' representation. Inevitably, because the number of state variables is usually less in this representation, this represents a more coarse level of resolution. Information has been lost in various ways. Characteristics are usually defined according to classes: age groups 0–4 years, 5–9, 10–14 etc. (though this can happen in the individual representation also). But more importantly, components are counted according to a subset of characteristics. For example, K_{pq} involved three, T^q_{ij} another three. But these two sets of variables do *not* tell us the (p, r), or age-income, breakdown of people by occupation, residence zone and job zone (or, conversely, the (i, j) breakdown of the (p, q, r) group). Complete information on the five characteristics (let alone the possibility of fifty) would need a five dimensional array such as X^{pqr}_{ij}. This will usually have too many 'cells' for practical purposes. Thus the model builder has to choose a number of arrays of smaller dimension and hope to capture the essential features of the system for his purposes even though information is being lost. This is why much system modelling is as much an act as a science!

Since indexed variables will play such a crucial role in the rest of this book, it is worthwhile to explore them in more depth. It turns out that they themselves contain much of the complexity of the models and that the student who can become at ease with such representations can make much progress using mathematical methods which in other respects are elementary.

The first and most essential point to note is that it is nearly always straightforward to 'translate' indexed state variables back into something concrete. Consider T^q_{ij}, for example. (Note, incidentally, that there is no fundamental significance in indices being represented as subscripts or superscripts — or occasionally in brackets after the main variable name. It is a matter of convenience and various methods are used in combination to avoid long single lists of indicators.) Zones i and j are 'typical' zones of a spatial system; the zones are exclusive and exhaustive regions of a map. Zones are usually numbered sequentially from 1 (up to N say) as shown in Figure 3.1. Their boundaries may be based on a grid, on administrative units such as wards, or hexagons, or whatever are appropriate and convenient. The variables then only have specific meaning when the letters are given specific numerical values. Thus, $T_{6,9}^3$ is the number of people in occupation group 3 who live in zone 6 and work in zone 9. This has a very clear and straightforward meaning. The use of letters as indices gives two advantages, each related to the *higher level of abstraction* thus achieved: first, it is possible to refer to trips from i to j for any i and any j—that is for typical values of the indices, without having to be more specific; secondly, the indices define a whole array of variables, sometimes written $\{T_{ij}^q\}$ and thus guarantees a comprehensive approach within which interdependences can be identified. The notation itself goes a long way to ensuring that all things are counted; nothing is neglected.

A two-dimensional array is a matrix and can be written out in standard form as

$$\{T_{ij}\} = \begin{bmatrix} T_{11} & T_{12} \ldots T_{1N} \\ T_{21} & T_{22} \ldots T_{2N} \\ \vdots \\ T_{N1} & T_{N2} \ldots T_{NN} \end{bmatrix} \qquad (3.1)$$

for an $N \times N$ array $\{T_{ij}\}$. Such an array is denoted here by a single letter in bold print, such as **T**. A three-dimensional array, such as $\{T_{ij}^q\}$, could be written as

$q = 1$:

$$\begin{bmatrix} T_{11}^1 & T_{12}^1 \ldots T_{1N}^1 \\ T_{21}^1 & T_{22}^1 \ldots T_{2N}^1 \\ \vdots \\ T_{N1}^1 & T_{N2}^1 \ldots T_{NN}^1 \end{bmatrix}$$

$q = 2$:

$$\begin{bmatrix} T_{11}^2 & T_{12}^2 \ldots T_{1N}^2 \\ T_{21}^2 & T_{22}^2 \ldots T_{2N}^2 \\ \vdots \\ T_{N1}^2 & T_{N2}^2 \ldots T_{NN}^2 \end{bmatrix}$$

$\cdots$

$q = Q$ (say):

$$\begin{bmatrix} T_{11}^Q & T_{12}^Q \ldots T_{1N}^Q \\ T_{21}^Q & T_{22}^Q \ldots T_{2N}^Q \\ \vdots \\ T_{N1}^Q & T_{N2}^Q \ldots T_{NN}^Q \end{bmatrix}$$

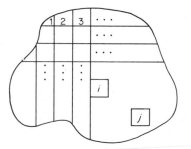

Figure 3.1 A spatial zonng system

Any systematic way of writing out the elements would do. Note also the advantage of using variables such as N and Q for the end of lists, which need be given specific values only in specific cases. Again, this is an example of variable names providing a higher level of abstraction and therefore, in this case, an array of variables which can be applied to *any* city. In each particular case, N and Q will be given particular values.

Table 3.1 *Examples of variables* (t labels, meaning 'at time t', excluded)

A_j	Total land area in zone j
A_j^U	Unusable land in zone j
A_j^B	Land used by the basic sector in zone j
A_j^R	Retail land use in zone j
A_j^H	Land available for housing in zone j
E_j	Total employment in zone j
E_j^B	Basic employment in zone j
E_j^{Rg}	Retail employment producing good g in zone j
T_{ij}	Number of residents of zone i working in zone j
W_i^{res}	Residential attractiveness of zone i
c_{ij}	Travel cost between zone i and zone j
β^{res}	Residence–work travel parameter
B_j	Balancing factor for residential model
P_i	Population of zone i
P_i^B	Basic population of zone i
f_j	Inverse activity rate in zone j
Z_1	Set of zones infringing residential density constraints
Z_i^H	Maximum residential density in zone i
Z_2	Set of zones not infringing residential density constraints
A_i^R	Balancing factor in residential model for zones infringing density constraints
S_{ij}^{HRg}	Flow of good g from zone j to residents of zone i
A_i^{HRg}	Balancing factor for home to centre model for good g
e_i^{Hg}	Expenditure per capita on good g by residents of zone i
W_j^{HRg}	Attractiveness of zone j for residents seeking good g
β^{HRg}	Travel parameter of residence to centre, good g, model
S_{ij}^{WRg}	Flow of good g from zone j to workers in zone i
A_i^{WRg}	Balancing factor for work to centre model for good g
e_i^{Wg}	Per capita expenditure on good g by workers in zone i
W_j^{WRg}	Attractiveness of zone j for workers seeking good g
β^{WRg}	Travel parameter of work to centre, good g, model
Z_3^g	Set of zones infringing minimum size constraints for good g
Z_4^g	Set of zones not infringing minimum size constraints for good g
$Z^{g\,min}$	Minimum employment, good g

We have now reached the point where we could organize any data we had on numbers in occupation groups by residential and workplace zones in an array (or table) of the form (3.2), and it is often necessary to do this. However, in a theoretical discussion, we can speak of the array $\{T_{ij}^q\}$ so that the notation in this case generates much economy and, more importantly, allows us to go on to deal with greater complexity.

Finally, note that the definition of state variables in this form relates closely to our earlier discussion of level of resolution. The level of spatial resolution is determined by the sizes of the zones in maps such as Figure 3.1 — the smaller the size the finer the resolution of course. Then, sectoral resolution is determined by the number of characteristics chosen to describe groups of components, and the 'widths' of the class intervals. Thus a scheme with 10 occupation groups would be a finer scale than one with 3.

The main subsystems of a city were shown in Figure 1.9. Table 3.1 above shows a set of state variables which describe the main features of most of these subsystems.

In subsequent chapters, a range of examples, mostly developed from those introduced in Chapter 1, will be used to illustrate the various methods. Variables of the type discussed here will be introduced and defined at appropriate points in the text. A reader who has any difficulty at such points should refer back to the subsection and reassure himself that the underlying principles involved with arrays of subscripted variables are straightforward.

NOTES ON FURTHER READING

The paper which is usually credited with having introduced the term 'quantitative revolution' into geography is by †Burton (1963). The mathematical equipment needed for the rest of this book is in †Wilson and Kirkby (1975, Second edition 1980 especially Chapters 1–3).

CHAPTER 4

Disorganized complexity 1: an introduction to entropy maximizing methods

4.1 INTRODUCTION

The concept of entropy as used in environmental modelling is introduced in the context of a specific example in Section 4.2. This concentrates on 'maximum entropy' as 'maximum probability'. The resulting model turns out to be one member of a family of spatial interaction models, the structure of which is outlined in Section 4.3. Most models involve no computational problems, but the tasks of finding balancing factors iteratively in the doubly constrained model, and of finding the decay parameters, β, by calibration, are discussed in Section 4.4. The concept of accessibility is also introduced at this stage. The principles underlying the building of entropy maximizing models are drawn together and summarized in Section 4.5 and a range of uses presented in Section 4.6. A water-resource example is presented in Section 4.7. The systems analytical aspects of entropy maximizing are reviewed in Section 4.8 and there is a concluding discussion on the range of interpretations of the entropy concept in Section 4.9.

4.2 ENTROPY AS PROBABILITY: THE JOURNEY TO WORK IN CITIES

The basic principles of entropy maximizing methods are explained here using the journey to work in cities as an example. It is tempting to use another example because this one has been used so much in the past. However, there remain a number of good reasons for sticking to it as an introduction: first, it is a system which is close to everyday experience which should help the reader to maintain a concrete vision of the modelling task; secondly, the notation used is a useful preliminary to the introduction of the family of models in the next section; and thirdly, if the reader does have any difficulties, there are alternative presentations in the literature which should be consulted — this being a useful way, for students especially, of tackling new and difficult concepts. However, it should be emphasized that the principles of the method are not difficult in spite of the mystique which sometimes seems to surround it!

54

Table 4.1

T_{ij}	Number of trips from zone i to zone j
O_i	Number of trip origins in zone i
D_j	Number of trip destinations in zone j
c_{ij}	Travel cost, in suitable units and components, from zone i to zone j
C	Total expenditure on travel for the journey to work
N	Number of zones

To fix ideas, consider the journey to work in a city in the morning peak period. Assume that the city is divided into a suitable zoning system in the usual way. The main state variables are those defined in Table 4.1. The level of resolution is relatively coarse at this stage, with no person types or transport modes identified for example.

As in Section 3.4, it is useful to define the variables for a general system of N zones. To fix ideas again, however, since this is the first major example, we can investigate the variable arrays and some of the relationships between them for a three-zone case: that is, $N = 3$. The trip array, $\{T_{ij}\}$, can be written out in full as shown in Figure 4.1. Such a table, in this particular case, is often called an origin–destination matrix. Note that the elements in the first row have a common first index and make up all the trips leaving zone 1 and so must sum to O_1:

$$T_{11} + T_{12} + T_{13} = \sum_{j=1}^{3} T_{1j} = O_1. \tag{4.1}$$

Similarly, the elements of the first column have a common second index and make up all the trips arriving at (having destinations at) zone 1 and so must add up to D_1:

$$T_{11} + T_{21} + T_{31} = \sum_{i=1}^{3} T_{i1} = D_1. \tag{4.2}$$

The sum of the second and third rows and the second and third columns are similarly interpreted and are all shown on Figure 4.1. The nine flows are shown

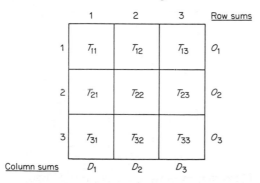

Figure 4.1   An origin–destination matrix

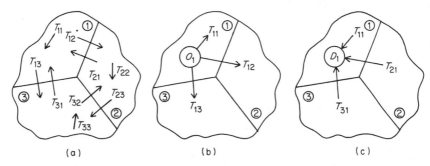

Figure 4.2 Spatial flows in a three zone system: (a) flows; (b) origin totals; (c) destination totals

on a map in Figure 4.2(a), and the first row and column sums in Figures 4.2(b) and 4.2(c) respectively. Row or column sums are sometimes called 'trip-ends'.

A set of *data* relating to these variables would take the form shown in Figure 4.3. That is, $T_{11} = 53$, $T_{12} = 27$, $T_{13} = 33$, $O_1 = 113$, and so on, for this particular case. Zone 1 is seen to be the dominant workplace zone, with zones 2 and 3 the main residential zones.

The $\{c_{ij}\}$ array can also be written out in full as shown in Figure 4.4(a) with some data shown (in time units, say) in Figure 4.4(b). The data in this case suggest that the best transport routes have been oriented towards zone 1 while the connections between zones 2 and 3, and within those zones, are relatively poor. The remaining variable, C, is related to the others:

$$C = \sum_{i=1}^{3} \sum_{j=1}^{3} T_{ij}c_{ij} = T_{11}c_{11} + T_{12}c_{12} + T_{13}c_{13} + T_{21}c_{21} + T_{22}c_{22}$$

$$+ T_{23}c_{23} + T_{31}c_{31} + T_{32}c_{32} + T_{33}c_{33}$$

$$= 47\,765 \tag{4.3}$$

for the case shown in Figure 4.4(b).

	1	2	3	Row sums
1	53	27	33	113
2	1500	40	200	1740
3	2000	350	51	2401
Column sums	3553	417	204	

Figure 4.3 An example of an origin–destination matrix

c_{11}	c_{12}	c_{13}
c_{21}	c_{22}	c_{23}
c_{31}	c_{32}	c_{33}

(a)

5	10	10
10	10	20
10	20	10

(b)

Figure 4.4 A cost matrix, with an example

Henceforth, we will use N-dimensional arrays (or $N \times N$ in the $\{T_{ij}\}$ and $\{c_{ij}\}$ cases). These can be shown diagrammatically as in Figure 4.5. In this case, the row and column sums can be exhibited for the ith row and jth column (as typical row and columns) and asserted to hold for $i = 1, 2, \ldots, N, j = 1, 2, \ldots, N$:

$$\sum_{j=1}^{N} T_{ij} = O_i, \qquad i = 1, 2, \ldots, N \tag{4.4}$$

$$\sum_{i=1}^{N} T_{ij} = D_j, \qquad j = 1, 2, \ldots, N \tag{4.5}$$

					Row sums
	T_{11}	T_{12}	. . .	T_{1N}	O_1
	T_{21}	T_{22}	. . .	T_{2N}	O_2
	. . .	. . .		. . .	. . .
	T_{N1}	T_{N2}		T_{NN}	O_N
Column sums	D_1	D_2	. . . . . .	D_N	

Figure 4.5 An $N \times N$ origin–destination matrix

and the total travel cost is given by

$$\sum_{i=1}^{N} \sum_{j=1}^{N} T_{ij} c_{ij} = C. \tag{4.6}$$

When the range of summation is clear, an equation such as (4.4) will be written as

$$\sum_j T_{ij} = O_i, \qquad \text{for each } i. \tag{4.7}$$

Note the power of the notation in this more general case. In the three-zone case, without picking out the ith row and jth column, six equations of the form of (4.1) and (4.2) would be needed. In the general case, 2 lines of algebra, in (4.4) and (4.5), represent N equations each, $2N$ in all. Even if $N = 1000$, the 2000 equations representing the row and column sums are still represented by (4.4) and (4.5) with the additional information that $N = 1000$.

The systems modelling task for this particular example is to represent the array of trip variables $\{T_{ij}\}$ as a function of the other variables $\{O_i\}$, $\{D_j\}$, $\{c_{ij}\}$, and C. In terms of Figure 1.9, $\{T_{ij}\}$ are the endogenous variables of the transport subsystem while the others are exogenous to this system (and assumed given for present purposes) but in fact relate to other subsystems: $\{O_i\}$ to residential location, $\{D_j\}$ to job location, $\{c_{ij}\}$ or $\{C\}$ to transport infrastructure and family expenditure respectively. Formally, we can write:

$$T_{ij} = T_{ij}(\{O_i\}, \{D_j\}, \{c_{ij}\}, C) \tag{4.8}$$

which simply means that T_{ij} is a function of all the variables shown in brackets (and the use of the array symbols there means that T_{ij} is in principle a function of $O_1, O_2, \ldots, O_N$ and not just O_i, and so on).

The scene is now set to describe the principle of entropy maximizing modelling in relation to this example. The argument turns on the definition of three levels of resolution and the relationships between them. They are termed macro, meso, and micro and are set out in diagrammatic form in Figure 4.6. At the macro level, it is assumed that the system state is described by the row and column totals and total travel cost and, indeed, that these quantities are exogenously given. At the meso level, the state variables are $\{T_{ij}\}$ (which in a sense include the macro level variables through equations (4.4)–(4.6)). The micro level is an individual-component representation: if individual A travels from zone 1 to zone 3, then his name is written into the $(1, 3)$ cell of the origin–destination table. So, at this level, it is necessary to imagine each cell of the table containing a list of names.

The state descriptions at the three levels are hierarchically related, as indicated on Figure 4.6(b). This is because any particular meso state can be generated from various different micro states; that is, by arranging the names in boxes in different ways (but such that the same $\{T_{ij}\}$ is produced for each alternative arrangement). A given macro state is generated by a number of possible meso states; this involves juggling the numbers within $\{T_{ij}\}$ but in such a way that the totals $\{O_i\}$, $\{D_j\}$, and C remain the same. The reader can check all this by experimenting with examples.

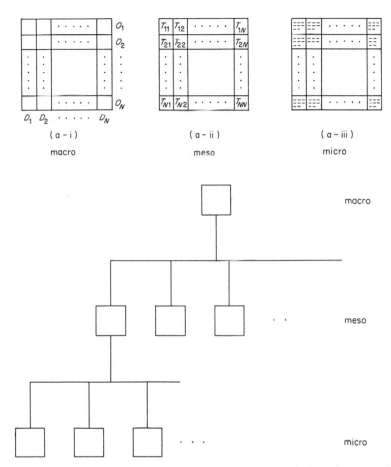

Figure 4.6 Micro, meso, and macro level state descriptions for a spatial interaction system

The meso state variables $\{T_{ij}\}$ are the ones we want to predict in a model. These are to be functions of *given* macro state variables $\{O_i\}$, $\{D_j\}$, and C. $\{c_{ij}\}$ can also, for present purposes, be assumed given. The key assumption of the entropy maximizing method can now be stated: *all micro states are equally probable*. This means that the probability of a meso state occurring is proportional to the number of micro states which give rise to it. This count of micro states, in other words, is a measure of the probability of a particular $\{T_{ij}\}$ occurring. Conceptually, therefore, the model building principle involves specifying a given macro state (as at the top of Figure 4.6(b)), which in effect represents the environment of the particular subsystem (in this case the transport system), identifying all possible micro states (as at the foot of the figure), and then finding the meso state which has the greatest number of micro states associated with it. Fortunately, the most probable state turns out to be very much the most probable.

The principle can be stated in mathematical terms as follows. First, define $W(\{T_{ij}\})$ to be the total number of micro states which give rise to $\{T_{ij}\}$. This can be calculated, algebraically, as a function of all the T_{ij}'s, in fact as*:

$$W(\{T_{ij}\}) = \frac{T!}{T_{11}!T_{12}!\ldots T_{NN}!} = \frac{T!}{\prod_{ij} T_{ij}!}. \tag{4.9}$$

So the mathematical problem is†

$$\text{Maximize} \quad \log W(\{T_{ij}\}) \tag{4.9a}$$

such that

$$\sum_{j} T_{ij} = O_i \tag{4.10}$$

$$\sum_{i} T_{ij} = D_j \tag{4.11}$$

$$\sum_{ij} T_{ij} c_{ij} = C. \tag{4.12}$$

The last three equations are the same as (4.4)–(4.6) and are repeated here for convenience. This problem can be solved using the method of Lagrangian multipliers (see Wilson, 1970, for the details), to give

$$T_{ij} = A_i B_j O_i D_j e^{-\beta c_{ij}}, \quad \begin{matrix} i = 1, 2, \ldots, N \\ j = 1, 2, \ldots, N \end{matrix} \tag{4.13}$$

where A_i and B_j are rather complicated factors given by

$$A_i = \frac{1}{\sum_j B_j D_j e^{-\beta c_{ij}}}, \quad i = 1, 2, \ldots, N \tag{4.14}$$

$$B_j = \frac{1}{\sum_i A_i O_i e^{-\beta c_{ij}}}, \quad j = 1, 2, \ldots, N. \tag{4.15}$$

The first important point to establish at this stage is that each T_{ij} can be estimated as a function of $\{O_i\}$, $\{D_j\}$, and $\{c_{ij}\}$. C appears implicitly through the parameter β (in the negative exponential function, $e^{-\beta c_{ij}}$). As C increases, β decreases, and vice versa. Secondly, the quantity

$$S = k \log W \tag{4.16}$$

---

*The exclamation mark here is the factorial sign: $N! = N(N-1)(N-2)\ldots 3.2.1$, e.g. $5! = 5 \times 4 \times 3 \times 2 \times 1$. $\prod$ is the product sign—like the summation sign but implying multiplication rather than addition.

†It turns out to be convenient to maximize $\log W$ rather than W, but this does not change the answer obtained.

where W is given by equation (4.9) turns out to be a measure of *entropy*: hence the name of the method. Further points are picked up in somewhat broader contexts at various places below: the model is discussed in its historical perspective at the beginning of the next section (4.3) as a preliminary to the introduction of a family of spatial interaction models. Computational aspects are discussed in Section 4.4. This particular approach to the concept of entropy and its relationship to disorganized complexity is compared to others in the final section (4.8). Finally, the mathematical aspects of the proof of (4.13)–(4.15) are discussed further in Chapter 6 since the problem is a nonlinear programming problem.

4.3 FAMILIES OF SPATIAL INTERACTION MODELS

It will already be clear that the journey to work is but one of many interaction and flow phenomena of interest in the environmental sciences. However, although the principles for building models are the same in many cases, there are important differences of detail. These are explored in this section and we will see that it is possible to develop a family of models and to learn how to choose specific members for particular purposes.

We will continue to use arrays of variables such as $\{T_{ij}\}$, $\{O_i\}$, $\{D_j\}$ and $\{c_{ij}\}$ though now bearing in mind that the variables may not necessarily be those connected to the journey to work. It is then a useful preliminary to give a very brief sketch of the history of spatial interaction model development and, in particular, of the so-called gravity model. A detailed history will not be given here (since it exists elsewhere—see Carrothers, 1956; Olsson, 1965; Wilson, 1970, 1974) but at least a view of the most elementary model puts the 'family' into perspective.

The gravity model arises from an analogy with Newtonian physics as depicted by Figure 4.7. The O_i and D_j terms are the 'masses' (equivalent to m_1 and m_2) and T_{ij} is the 'force' (equivalent to F_{12}). The distance d_{12} is replaced by the travel cost, or other measures of impedance, c_{ij}. Newton's inverse square law is

$$F_{12} = G\frac{m_1m_2}{d_{12}^2} \tag{4.17}$$

for some suitable constant G. An exactly equivalent geographical law would be

$$T_{ij} = K\frac{O_iD_j}{c_{ij}^2} \tag{4.18}$$

for some constant, K. A square law is rather arbitrary in a social context, and so

Figure 4.7 Variables for gravity and spatial interaction models

the '2' is replaced by a parameter, say n, to give

$$T_{ij} = K \frac{O_i D_j}{c_{ij}^n} = K O_i D_j c_{ij}^{-n}. \tag{4.19}$$

This model has a family resemblance to (4.13) which was derived using the entropy maximizing method. This can be seen most clearly if we write:

$$T_{ij} = (\text{factors})(i\text{-mass term})(j\text{-mass term})$$
$$\times (\text{decreasing function of impedance}). \tag{4.20}$$

The entropy maximizing model has factors $A_i B_j$ instead of simply K. The 'mass' terms are the same. The gravity model has a power function as impedance function while the entropy model has negative exponential. The different factors are important and play a role in defining differences within a family below; the different impedance functions are less important, though the fact that the negative exponential function arises naturally in the entropy maximizing method forms the basis of a useful interpretation which will be pursued below (in Section 4.5).

Why are the differences in 'factors' so important? A little experimentation with a three-zone numerical example will soon convince the reader that it is impossible to choose K in equation (4.19) in such a way as to ensure that equations (4.10) and (4.11) are satisfied — that is, to ensure that the trip end totals derived from the $\{T_{ij}\}$ variables predicted by the model are what they were originally assumed to be. In the journey to work example, it was reasonable to assume that $\{O_i\}$ and $\{D_j\}$ were known. In other cases this may not be true, as examples will demonstrate shortly. Four distinct cases can be distinguished: (i) neither set of trip ends, $\{O_i\}$ or $\{D_j\}$, is known; (ii) origin trip ends $\{O_i\}$ (sometimes called *productions*) are known, but not $\{D_j\}$; (iii) destination trip ends, $\{D_j\}$ (sometimes called *attractions*) are known, but not $\{O_i\}$; (iv) both $\{O_i\}$ and $\{D_j\}$ are known. If the origin trip ends are known, equation (4.10) must hold (which is repeated here for convenience):

$$\sum_j T_{ij} = O_i, \qquad i = 1, 2, \ldots, N \tag{4.21}$$

and these equations are known as *production constraints* (since the $\{T_{ij}\}$ are to be *constrained* to sum to the appropriate row totals). If $\{D_j\}$ variables are known, then

$$\sum_i T_{ij} = D_j, \qquad j = 1, 2, \ldots, N \tag{4.22}$$

and these are known as *attraction constraints*. In the fourth case, of course, both sets of constraints must hold simultaneously.

The four cases are known as unconstrained, production-constrained, attraction-constrained (the last two are examples of singly constrained models), and production-attraction (or doubly) constrained. An appropriate model can be derived using entropy maximizing principles in each case. The key to the method is to include as trip end constraint equations whichever ones relate to known information: (i) neither of the equations (4.21) or (4.22); (ii) (4.21); (iii) (4.22); (iv) (4.21) and (4.22). In each case the model takes the form of equation (4.20) and the factors are defined and constructed to ensure that the appropriate constraints are satisfied. The mass terms are O_i or D_j if assumed known; or 'attractiveness factors', $W_i^{(1)}$ or $W_j^{(2)}$ associated with origins or destinations respectively where trip ends are not known. Thus, if we use the negative exponential distance function, the main model equations are:

(i) *unconstrained*

$$T_{ij} = KW_i^{(1)}W_j^{(2)}e^{-\beta c_{ij}} \tag{4.23}$$

(ii) *production-constrained*

$$T_{ij} = A_iO_iW_j^{(2)}e^{-\beta c_{ij}} \tag{4.24}$$

(iii) *attraction-constrained*

$$T_{ij} = B_jW_i^{(1)}D_je^{-\beta c_{ij}} \tag{4.25}$$

(iv) *production-attraction constrained*

$$T_{ij} = A_iB_jO_iD_je^{-\beta c_{ij}} \tag{4.26}$$

The factors are calculated to ensure that the trip flows sum to the assumed-known trip ends (and hence they are sometimes known as *balancing factors* because they 'balance' the matrix). In the unconstrained case, some rule has to be adopted for calculating K, and it is customary to assume that the total flow

$$T = \sum_i \sum_j T_{ij} \tag{4.27}$$

is known and K is calculated to satisfy this equation. Thus, the balancing factors are

(i) *unconstrained*

$$K = \frac{T}{\sum_i \sum_j W_i^{(1)}W_j^{(2)}e^{-\beta c_{ij}}} \tag{4.28}$$

(ii) *production-constrained*

$$A_i = \frac{1}{\sum_j W_j^{(2)}e^{-\beta c_{ij}}} \tag{4.29}$$

(iii) *attraction-constrained*

$$B_j = \frac{1}{\sum_i W_i^{(1)}e^{-\beta c_{ij}}} \tag{4.30}$$

(iv)   *production-attraction constrained*

$$A_i = \frac{1}{\sum_j B_j D_j e^{-\beta c_{ij}}}$$

(4.31)

$$B_j = \frac{1}{\sum_i A_i O_i e^{-\beta c_{ij}}}$$

(4.32)

and the complete models are given by one of equations (4.23)–(4.26) with the appropriate balancing factor equations. In the singly-constrained case, the balancing factor expression is sometimes substituted directly into the main model equation. For example, in the production-constrained case, we can substitute A_i in (4.29) into (4.24) to give

$$T_{ij} = \frac{O_i W_j^{(2)} e^{-\beta c_{ij}}}{\sum_j W_j^{(2)} e^{-\beta c_{ij}}}.$$

(4.33)

This is a useful form of the model for some purposes (in the sense that interpretation is easier — the two forms are mathematically equivalent of course).

We now see that the journey to work model is an example of a doubly constrained model and that the original gravity model, in equation (4.19), is an unconstrained model (though with a power function instead of a negative exponential function) and that the mass terms should be attractiveness terms rather than trip end totals.

In effect, the family of models presented here forms a model building kit — a set of rules which allows the appropriate model to be constructed for each of a wide range of examples. This includes identifying the type of phenomenon involved, defining (or using existing) appropriate variables, and then constructing the model of the appropriate type. These principles can now be illustrated with basic examples of each type. A number of refinements will be considered later in Section 4.5.

The unconstrained case is relatively rare. One example is the allocation of people to homes (i) and workplaces (j) (T_{ij}, say) in a city they have migrated to. Suppose there are Q such people in all, and totals of H_i homes in each zone i, E_j jobs in each zone j, and the usual impedance matrix $\{c_{ij}\}$. In this case, we would have $Q \ll \sum_i H_i, \sum_j E_j$, (where '$\ll$' means 'very much less than') so the H_i's and Q_j's are acting as attractiveness factors K and therefore play the role of $W_i^{(1)}$, $W_j^{(2)}$ in equation (4.23). The model is thus:

$$T_{ij} = KH_i E_j e^{-\beta c_{ij}}$$

(4.34)

with

$$K = \frac{Q}{\sum_i \sum_j H_i E_j e^{-\beta c_{ij}}}.$$

(4.35)

The production-constrained case is illustrated by the well-known Huff (1964), Lakshmanan-Hansen (1965) shopping model. The interaction term is S_{ij}, the flow of cash from residents of zone i to shops in zone j. This is production constrained because if e_i is the average per capita expenditure by residents of zone i, P_i is the population of zone i, then $e_i P_i$ is the amount of cash available, is known, and plays the role of O_i. The attractiveness of shops in zone j is initially taken as shopping centre size (measured as floorspace), say W_j, raised to some power to represent scale economies. That is, $W_j^{(2)} = W_j^{\alpha}$. If $\{c_{ij}\}$ is the usual impedance matrix, the model is

$$S_{ij} = A_i(e_i P_i) W_j^{\alpha} e^{-\beta c_{ij}} \qquad (4.36)$$

where

$$A_i = \frac{1}{\sum_j W_j^{\alpha} e^{-\beta c_{ij}}} \qquad (4.37)$$

which ensures that

$$\sum_j S_{ij} = e_i P_i. \qquad (4.38)$$

This example enables another general point to be illustrated: that models which are not fully constrained also act as *location* models as well as interaction models, which is obviously of substantial geographical significance. In this case, the column sums of the interaction matrix, $\sum_i S_{ij}$ represent the total use of shopping centres. The notation S_{*j}, with an asterisk replacing an index denoting summation, is also sometimes useful. Thus, a prediction of the locational distribution of shopping centre usage is generated by the model.

The attraction-constrained model is, of course, the mirror image of the other. A further example may be useful, however. Consider a residential location model based on a given distribution of jobs (E_j) and the distribution of workers to residences around workplaces on the basis of attractiveness factors $\{W_i\}$ and the usual impedance matrix $\{c_{ij}\}$. If the interaction variable is T_{ij}, then

$$T_{ij} = B_j W_i E_j e^{-\beta c_{ij}} \qquad (4.39)$$

where

$$B_j = \frac{1}{\sum_i W_i e^{-\beta c_{ij}}} \qquad (4.40)$$

to ensure that

$$\sum_i T_{ij} = E_j. \qquad (4.41)$$

The main purpose of this model is to calculate the residential distribution T_{i*} given by

$$T_{i*} = \sum_j T_{ij}. \qquad (4.42)$$

The doubly-constrained model has already been illustrated by the journey to work example.

So far, one basic family of spatial interaction models has been presented and illustrated. This section was titled 'families' because others can be developed too. They arise out of the addition of more complexity to the basic models as is obviously necessary in many cases to simulate reality in an effective way. For example, models often need to be disaggregated: journey to work by transport mode (k) and person type $(n$, perhaps denoting car owner/non car owner) lead to a model of the form

$$T_{ij}^{kn} = A_i^n B_j O_i^n D_j e^{-\beta^n c_{ij}^k} \tag{4.43}$$

where the variables have obvious definitions building on the earlier, more aggregated, ones. In this case, the balancing factors are

$$A_i^n = \sum_{jk} B_j D_j e^{-\beta^n c_{ij}^k} \tag{4.44}$$

to ensure that

$$\sum_{jk} T_{ij}^{kn} = O_i^n \tag{4.45}$$

and

$$B_j = \sum_{ikn} A_i^n O_i^n e^{-\beta^n c_{ij}^k} \tag{4.46}$$

to ensure that

$$\sum_{ikn} T_{ij}^{kn} = D_j. \tag{4.47}$$

This is obviously a member of a broader family of models than the equivalent aggregate one.

Further elaboration at this stage would take us beyond the scope of this introductory book and the reader is referred to the more advanced literature (such as Wilson, 1974).

4.4 COMPUTATIONAL ASPECTS

So far, we have concentrated on basic concepts rather than numerical work with the models. It is essential to full understanding that numerical and practical experience is gained, however. For any but the simplest examples, this involves the use of electronic computers. Only a 3×3 or perhaps at most a 4×4 example could be calculated using a desk machine. The main purpose of this section, beyond encouraging the reader to carry out some numerical work, is to add light to the conceptual base by noting some computational points which may at first seem odd to the uninitiated.

There are essentially no computational problems associated with unconstrained or singly constrained models. Numerical values can be inserted for variables on the right-hand side of equations and the variables on the left-hand side can then be calculated. A standard computer programme is often near at hand, certainly in universities or planning offices. The perceptive reader will, however, have noticed a different kind of problem with the doubly constrained model, say that given by equations (4.26), (4.31), and (4.32), which are repeated here for convenience:

$$T_{ij} = A_i B_j O_i D_j e^{-\beta c_{ij}} \tag{4.48}$$

$$A_i = \frac{1}{\sum_j B_j D_j e^{-\beta c_{ij}}} \tag{4.49}$$

$$B_j = \frac{1}{\sum_i A_i O_i e^{-\beta c_{ij}}}. \tag{4.50}$$

Each A_i in (4.49) is a function of all the B_j's; but each B_j in (4.50) is a function of all the A_i's! Such equations have to be solved iteratively. For example, initially, all the B_j's could be set to 1, the A_i's calculated from (4.49), new B_j's from (4.50), new A_i's from (4.49), and so on. Fortunately, such a process usually converges to an acceptably accurate answer rapidly. If we allow q to label iteration numbers, this process can be written:

$$B_j^{(1)} = 1 \tag{4.51}$$

$$A_i^{(q)} = \frac{1}{\sum_j B_j^{(q)} D_j e^{-\beta c_{ij}}} \tag{4.52}$$

$$B_j^{(q+1)} = \frac{1}{\sum_i A_i^{(q)} O_i e^{-\beta c_{ij}}} \tag{4.53}$$

and we would cycle round (4.52) and (4.53) putting $q = 1, 2, \ldots, Q$, where Q would be the iteration at which acceptable accuracy was achieved. This would be when,

$$\frac{|A_i^q - A_i^{q-1}|}{A_i^q} < \delta \qquad \text{for all } i \tag{4.54}$$

and

$$\frac{|B_j^{(q+1)} - B_j^{(q)}|}{B_j^q} < \delta \qquad \text{for all } j \tag{4.55}$$

where δ is a parameter — say 0.001, so that the iteration stops when the factors all change by less than a tenth of a percent from the previous step. These A_i and B_j values would then be inserted into (4.48) and T_{ij} calculated.

Another potential mystery arises from the question: 'where do parameters such as β come from?' They are obtained by *calibration*, which is the process of comparing $\{T_{ij}\}$ predicted by the model with a set of observations of the same array, perhaps from a sample survey, and denoted by $\{N_{ij}\}$, and finding the β

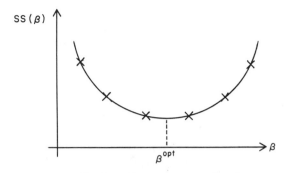

Figure 4.8 Best fit parameters estimation

which gives the best fit between predicted and observed. One measure of goodness of fit is the 'sum of squares', denoted say by SS:

$$SS = \sum_{ij} (T_{ij} - N_{ij})^2 \tag{4.56}$$

which is obviously zero for a perfect fit and increasingly large for bad fits. SS is, in effect, a function of β (since all the T_{ij}'s on the right-hand side depend on β — through equation (4.48) if we stick to this example for the present — and could be written SS(β). The model is run for a range of β values, and SS(β) calculated for each run. This can then be plotted against β and the plot may take the form shown in Figure 4.8. The 'best fit' β value is then shown as β^{opt}, the value for which SS(β) is a minimum. Calibration can be carried out exactly as described, and this illustrates the concept. In practice, however, there are powerful computer algorithms which remove this kind of hard labour. It should also be emphasized that there are other measures of goodness-of-fit, and that perhaps the best calibration method is the maximum likelihood method (and 'likelihood' turns out to have a close relationship to entropy as we will see in Section 4.7). This is based on the result mentioned earlier that β and C vary inversely. However, a discussion of these would take us beyond the scope of this book and the reader is referred elsewhere (see, for example, Batty, 1976).

4.5 PRINCIPLES FOR BUILDING ENTROPY MAXIMIZING MODELS: SUMMARY AND EXTENSIONS

At this stage, it is useful to underline the general principles which have been established by example for building entropy maximizing models. There are four main steps. *First*, ensure that the system of interest is of a type which is suitable for the entropy maximizing method. *Secondly*, define suitable state variables. The endogenous variables will be the ones to be predicted in this particular model. The exogenous variables will be connected to the endogenous variables implicitly via the constraints and explicitly via the model built. The *third* step is in fact the articulation of the constraints. These contain as much as possible of the model builder's *information* on interrelationships in the system. *Fourthly*, form an entropy function in terms of the endogenous variables and maximize this subject to the constraints.

The next point to note is that the model builder often has to work very hard to take his product nearer to reality; the preceding discussion perhaps gives an impression of ease and tidiness which is not justified. To illustrate the kinds of things which are involved, it is convenient to return to the journey to work and similar spatial interaction examples. We have already noted the need for disaggregation, for example to take as a basic state variable T_{ij}^{kn} as the flow of people of type n from zone i to zone j by mode k; or S_{ij}^{kg} as the flow of people from zone i to zone j by mode k to purchase goods of type g.

The mass terms usually look after themselves if they are 'known', for example O_i^n and D_j in equation (4.43). Care has to be taken with disaggregation, however. Note in this example that O_i becomes O_i^n, since origins can be identified by car owner/non car owner characteristics, but D_j can not be so subdivided: the two types of people are assumed to compete for the same jobs.

The mass terms are more complicated to deal with when they are attractiveness terms. Consider W_i in equation (4.39), for example. It is intended to reflect the attractiveness of zone i for housing relative to other zones. Such a concept involves many factors, such as availability of land (say L_i), an index of quality (say Q_i), an index of the social class of near neighbours (say S_i), and accessibility to jobs and shops (say X_i^J and X_i^S respectively). Then, we might take

$$W_i = L_i^{\alpha_1} Q_i^{\alpha_2} S_i^{\alpha_3} X_i^{J\alpha_4} X_i^{S\alpha_5}. \tag{4.57}$$

A great deal of empirical work is then needed to establish whether this is the most appropriate functional form for W_i, the relative values of parameters such as α_1–α_5, and the most effective way of measuring the individual factors (if indeed that is possible). Relatively little of this sort of work has been done.

This is a useful point at which to interject something on the measurement of *accessibility*, referred to above as a component of attractiveness. This concept has an intuitive meaning, perhaps several, and there are many ways in which any such meaning can be quantified. One particularly useful one, which originated with Hansen (1959), can be based on the spatial interaction model itself. The intuitive idea is this: accessibility, say from a residential zone i to jobs (X_i^J above) declines when the job opportunities are an increasing distance from i, but increases with the number of opportunities at a particular location. This can be expressed as:

$$X_i^J = \sum_j D_j e^{-\beta c_{ij}} \tag{4.58}$$

where β is the parameter of the journey to work model (or even the residential location model itself). In other words, the latter model provides terms which measure the way in which increasingly distant opportunities decrease the accessibility measure. Note, in the case of this and the preceding paragraph, the possibility of a term in a model (X_i^J) being defined in terms of variables and parameters (in this case β) which also appear in the model. It is obviously important to avoid circularity, but this can be done and the complicated relationship is in fact an example of feedback, as shown in Figure 4.9.

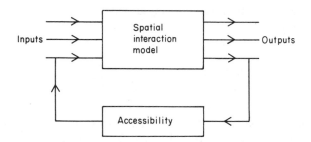

Figure 4.9 'Accessibility feedback' in a spatial interaction model

The spatial interaction model has a number of inputs (the exogenous variables) and produces a number of outputs (the endogenous variables). One of the outputs, accessibility, may also be an input as shown. A time lag is usually assumed to operate in such mechanisms: the value of accessibilities produced as an output when the model is run for time t, say, provides an input for the model run at time $t+1$.

The next term for detailed consideration is the impedance function. To be specific, consider $e^{-\beta^n c_{ij}^k}$ in equation (4.43). The first point to consider is the functional form: should it be negative exponential or something else? Again, this is a matter of detailed empirical investigation, but one interesting point arising out of the entropy maximizing theory can be noted. The negative exponential arises naturally in the theory. Suppose that empirical work shows a power function $(c_{ij}^k)^{-\beta^n}$ to give the best fit. Note that

$$(c_{ij}^k)^{-\beta^n} = e^{\log(c_{ij}^k)-\beta^n} = e^{-\beta^n \log c_{ij}^k}. \tag{4.59}$$

The power function, therefore, is behaving like a negative exponential function but with $\log c_{ij}^k$ as cost instead of c_{ij}^k. This can be interpreted as the people in this system perceiving cost in a log-like way. In distance terms, this is equivalent to saying that an additional mile following one mile is perceived as greater than an additional mile following a hundred miles. It is perhaps not surprising, therefore, that for spatial systems containing mainly short trips, the negative exponential function fits well, while for those with a mix of trips, but including plenty of long ones, the power function is often more effective. Thus, whether the negative exponential function fits or not helps with interpretation; it is not directly a test of whether entropy maximizing methods are successful or not.

Parameters β^n would be estimated using the methods sketched in the previous section. The remaining question, therefore, is how to measure c_{ij}^k. As with attractiveness terms, a variety of factors can contribute to what is usually called 'generalized' cost; for example, travel time (t_{ij}), money costs (taken as proportional to distance, d_{ij}), 'excess' time (waiting, parking, etc., e_{ij}) and so on. Such terms are usually combined linearly:

$$c_{ij} = a_1 t_{ij} + a_2 d_{ij} + a_3 e_{ij} + \dots . \tag{4.60}$$

Since the term $a_2 d_{ij}$ is measured in money terms (say petrol costs, or fares, per mile times distance), coefficients such as a_1 and a_3 provide money measures of different kinds of time—in itself an interesting idea. Quite a lot of empirical work has been carried out on this task. The remaining difficulty, which is a very serious one, is that travel time is a function of the amount of congestion on the underlying networks. This topic will be pursued further in Chapter 7, however.

4.6 SOME USES OF ENTROPY MAXIMIZING MODELS IN URBAN PLANNING

Suppose a model can be developed and calibrated effectively. What sort of uses does it have? This question will mainly be tackled in Chapter 10, when the full range of methods for particular planning problems will be available. However, it may be useful to give a preliminary sketch in relation to the models presented in this chapter.

The transport model is used to assess the impacts of alternative plans (designs) as shown in Figure 4.10. The plan parameters form some of the inputs to the transport model. Thus, for any plan, a set of impact measures are generated, and these form the inputs for the calculation of evaluation criteria. These, of course, are related to whatever goals and objectives society is seeking to achieve. Examples will be pursued in Chapters 9 and 10. In the light of what is then seen to be achieved from a particular plan, revised designs can be generated and tested as shown. In effect, the planner is able to experiment with his study area in his office by using the transport model to represent its enormous complexity.

Alternative plans may relate to the spatial distribution of land use (which would be reflected in the O_i's and D_j's), in the provision of transport networks or in traffic management policies (each of which would affect the c_{ij}^k's, the former in a very complicated way through the underlying network relationships mentioned earlier). The importance of having the model to handle complexity arises from the high degree of interdependence reflected in the model. Even quite a simple sounding plan, to change one O_i (a new housing estate) or one network link (a new section of road), affects every T_{ij}^{kn} element in the model. This interdependence arises mainly through the A_i and B_j factors. Some effects are more important than

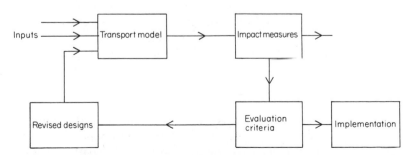

Figure 4.10 The transport model and design

others, of course, but these can be identified within the model and the fear removed that important effects are being forgotten or neglected.

So far, we have concentrated on the use of entropy-maximizing transport models, whether the plans involve the provision of network links or the rearrangement of land uses. We saw in Section 4.3, however, that singly-constrained spatial interaction models can be used to predict *locational* variables, as in the retail model cited as an example in equations (4.36)–(4.38). If this model is used, the locational variable to be calculated is total revenue at j, given all the other variables. If this is denoted by D_j, we have

$$D_j = \sum_i S_{ij}. \tag{4.61}$$

The attractiveness terms, W_j, are usually measured by 'size' and so are a measure of (possibly planned) supply. The model can thus be used as a basis of planning exercises within which alternative W_j-configurations are explored and the model offers the set of D_j's for each alternative plan. For each such plan, therefore, the viability of each centre can be examined.

4.7 AN APPLICATION TO A WATER RESOURCE SYSTEM

We have seen that a water resource system can be considered as a set of sources and sinks joined by the links of a distribution network. There are then some obvious similarities of structure between the transport system discussed in the previous section and the present one. O_i and D_j, for example, can be interpreted as supply and demand at sources and sinks respectively (and in this case, origin and destination zones will each be made up of different sets). There is also a common feature in that many links of the network will carry flows involving several origin–destination pairs. It is this feature which leads to a possible application of the entropy maximizing distribution model to the water resource system.

If a system is defined at a coarse scale, or with sources and sinks widely separated, then it may be possible to identify the flows from each source to each destination—and we use the notation $\{T_{ij}\}$, interpreted as flow per unit time, say—with some precision. In a dense network with many nodes, however, this may not be the case: because of the complexities of what happens at particular junctions, it is not easy to say what part of j's demands is being met by i. Yet as we will see in discussing optimization problems in Chapter 10, such knowledge is important in order to get an accurate attribution of costs.

Suppose the flows are described by a doubly-constrained spatial interaction model. We could write this as

$$T_{ij} = A_i B_j O_i D_j e^{-\beta d_{ij}} \tag{4.62}$$

with notation as in the transport case (but with terms suitably re-interpreted) except for d_{ij} replacing c_{ij} as we will assume the pattern of flow is governed by distance travelled. (In practice, as in the transport case, the pattern will be more complicated, because of the varying capacities of network links.) This model could be calibrated, using the methods of Section 4.4, if some knowledge of 'average distance travelled' was available. The measurement of such a quantity presents an interesting problem for hydrologists!

We return in Chapter 10 to a discussion of how this kind of information could be used.

4.8 SYSTEMS ANALYTICAL ASPECTS OF ENTROPY MAXIMIZING

A number of the basic concepts of systems theory which were introduced in Chapter 2 have already been relevant to the discussion of entropy maximizing methods presented so far. Others have been implicit. The purpose of this section is to examine the role of systems theory in entropy maximizing methodology more comprehensively.

The issues concerned with definition of *system, environment,* and *boundary* are all reasonably clear, but an interesting theoretical issue is raised at the outset as to whether systems of interest are *closed* or *open*. The entropy maximizing method (as is made crystal clear in its application in physics) of the type applied here makes closed system assumptions. However, this is not as restrictive as it sounds, since the exogenous variables, such as O_i, D_j, and C (which determines β) in the journey to work model example, describe the *environment* of the system; so that if the environment in that sense changes, then so do the exogenous variables and hence the model's prediction of the endogenous variables (the T_{ij}^{kn}'s in the example). This does highlight another assumption of the method: that, if there is a change, the system moves to its new *equilibrium* position very quickly, for the model is of an equilibrium state. Transient states are ignored (or, equivalently, relaxation times are assumed to be very short).

We saw that the concept of *level of resolution,* and the definition of the system state at three distinct levels, is important to the theory. *Couplings* between system components are weak. Indeed, they have to be in order for the 'Weaver-II-disorganized complexity' assumptions to hold. If these assumptions break down then the equal probability assumptions at the micro level also break down and the whole method is invalid. Nonetheless, there can be strong interdependence at the meso level, as we saw in the brief discussion on the use of the models in the previous section. Couplings are also important in the representation of connections between *subsystems,* mainly through common variables (endogenous in one subsystem, exogenous in others). The concepts of *input* and *output* are also particularly relevant here.

The concepts of *stocks* (O_i's and D_j's), *flows* (T_{ij}^{kn}'s), *channels, routes* through *networks* all have obvious roles to play in spatial interaction and location models. And we have seen at least two examples of *feedback* (within the model, in Figure 4.9, and in the use of the model, in Figure 4.10).

Most of the remaining concepts of Figure 2.17 are concerned with *dynamics* (*structure, behaviour, activity, phase space*) and will be considered further in Chapter 8. Others are concerned with planning (such as *optimizing principles*) or *control* and will be considered in Chapter 10.

4.9 THE ENTROPY CONCEPT: A CONCLUDING DISCUSSION

The concept of entropy presented in Section 4.2 was directly related to the probability of a particular state of the system occurring. This notion of entropy is that used in statistical mechanics in physics, though this does not mean that the use here relies on any analogy with physics: each is an example of a particular application of a higher level method which transcends disciplines defined in system of interest terms.

There are, however, other ways in which the concept can be built up, and the particular choice is often a matter of controversy. Most other concepts rely on the notion of the probability of system *components* being in a given state. To fix ideas, we continue to use the journey to work example. The probability of an individual living in i and working in j can then be defined in frequency terms as

$$p_{ij} = \frac{T_{ij}}{T}. \tag{4.63}$$

The *entropy of a probability distribution* can then be defined as

$$S = -\sum_{ij} p_{ij} \log p_{ij}. \tag{4.64}$$

If we take $k = 1$ in equation (4.16), then (4.9) and (4.16) imply that our earlier definition of entropy was

$$S = \log\left(\frac{T!}{\prod_{ij} T_{ij}!}\right). \tag{4.65}$$

This can be written

$$S = \log T! - \sum_{ij} \log T_{ij}!$$

$$= log\, T! - \sum_{ij} T_{ij} \log T_{ij} + \sum_{ij} T_{ij}$$

$$= \log T! + T - \sum_{ij} T_{ij} \log T_{ij}. \tag{4.66}$$

Using (4.63), equation (4.64) can be written

$$S = -\sum_{ij} \frac{T_{ij}}{T} \log \frac{T_{ij}}{T}$$

$$= -\frac{1}{T}\sum_{ij} T_{ij}(\log T_{ij} - \log T)$$

$$= -\frac{1}{T}\sum_{ij} T_{ij} \log T_{ij} + \log T \tag{4.67}$$

(after some manipulation). This shows that the two entropies are proportional, and this means that models obtained by maximizing either will be the same.

The expression (4.64) first occurred in *information theory* in the work of Shannon and Weaver (1949) and was interpreted by Jaynes (1957) as the amount of *uncertainty* associated with a probability distribution; conversely, $-S$ is a measure of the amount of information associated with it. This interpretation is consistent with the earlier one since the maximum entropy meso state is that with the greatest number of possible micro states associated with it — and hence the greatest degree of uncertainty at the micro scale. In Jaynes' representation, the constraints are seen as representing known information about the *expectation* values of certain quantities associated with the system. For example,

$$\sum_{ij} T_{ij}c_{ij} = C \tag{4.68}$$

could be written as

$$\sum_{ij} \frac{T_{ij}}{T} c_{ij} = \sum_{ij} p_{ij}c_{ij} = \frac{C}{T} = \bar{c} \tag{4.69}$$

and be interpreted as a statement of the known mean of $\{c_{ij}\}$. Entropy maximizing modelling is then interpreted as maximizing uncertainty subject to known information, and this means *minimizing bias on the part of the observer*. Any other model for p_{ij} (and hence T_{ij}) would involve the assumption of information which the observer, as analyst, by hypothesis is assumed not to have. If more information turns up, of course, then it can be incorporated into the procedure as additional constraints.

A useful development, which does lead to an extended modelling procedure to be discussed in another context in the next chapter, is that due to Kullback (1959). He defines

$$S = -\sum_{ij} p_{ij}\log\left(\frac{p_{ij}}{q_{ij}}\right) \tag{4.70}$$

to be a measure of the amount of uncertainty associated with a distribution $\{p_{ij}\}$ obtained by adding information, in the form of new constraints, to an original distribution $\{q_{ij}\}$. Equation (4.64) can be seen as a special case of this with $q_{ij} = 1$ for all i and j.

Yet another related development arises from statistics. There, a likelihood function is defined as

$$L = \prod_{ij} p_{ij}^{T_{ij}} \tag{4.71}$$

and since

$$\log L = \sum_{ij} T_{ij}\log p_{ij} \tag{4.72}$$

it is easy to see that this is proportional to *minus* the entropy function. At first sight, this seems rather odd since statisticians maximize log likelihood which would be equivalent to minimizing entropy. However, this puzzle can be resolved as follows. The entropy maximizer is finding the functional form of $\{p_{ij}\}$ and to do this by minimizing likelihood is to make the least biased assumption as before. The statistician is estimating the parameters of an assumed distribution. So once the functional form is determined maximum likelihood can be used in model calibration. It turns out to be a very neat procedure because it can be shown that if the constraint equations are solved for the parameters, then this provides a maximum likelihood estimate of these parameters.

The various concepts of entropy which have been introduced so far are, to this author's mind, all basically similar. It seems best to choose a formulation one is comfortable with and to stick to that, but to know enough about the other formulations to be able to transform if necessary. Occasionally this is especially useful when one formulation performs a task particularly well—as with the use of the Kullback measure when information is added. One final minor point can usefully be added here before we proceed to a brief discussion of some rather different uses of the entropy concept. That is, the measure discussed so far has always referred to discrete distributions with classes labelled $i, j, \ldots$ or whatever. It can also be applied to continuous distributions with entropy defined as

$$S = - \int p \log p \, \mathrm{d}p. \tag{4.73}$$

We now consider three different uses of the entropy concept. First, it is sometimes used as a statistical index, especially in ecology for example, as a measure of *diversity*. Suppose p_i is the proportion of species i found in some ecosystem. Then

$$S = - \sum_i p_i \log p_i \tag{4.74}$$

serves as an index of diversity which ranges between 0 and -1. If there are N species, the index takes minimum value of -1 when each $p_i = 1/N$, which indicates diversity. It takes its maximum value of 0 when one p_i is one and all the rest zero, which obviously indicates lack of diversity (see Pielou, 1967, for a detailed account).

Perhaps the most common popular view of entropy is as a concept associated with physics, and especially with the second law of thermodynamics. The entropy concept *is* basically the same, as can be seen from the derivation of the laws of thermodynamics via statistical mechanics, but it is used in thermodynamics to express properties of particular physical systems. That 'entropy always increases', for example, says two specific things about certain physical systems. First, if the system is in a *dis*equilibrium state, then it tends towards an equilibrium state represented by maximum entropy. Secondly, if the environment of the system is changed, but in a way which does not involve additional energy, then the system will tend to a new equilibrium state and entropy will increase. The second law of thermodynamics, therefore, says something about the direction of change in physical systems. The law is sometimes stated in rather dramatic form as 'disorganization always increases' or 'all systems tend to a state of chaos'. This is because highly ordered systems of gases have relatively low energy and therefore do not occur in nature. It is very important that the reader does not associate 'choas', or 'disorganization', except in the technical Weaver-II sense, with the states predicted by the entropy maximizing models of this chapter.

It is an interesting question, which has occasionally been pursued but not as yet with very striking conclusions, as to whether there are 'social' equivalents of the second law of thermodynamics. And it may be possible to develop a related concept for systems of organized complexity in relation to the way they maintain order: since some mechanism, and 'energy' supply, is needed for order to be maintained, this is sometimes described as the supply of negentropy, a quantity defined to be 'minus entropy'. This means, in effect, the supply of energy or something in a social system which plays the role of energy (like 'money'?), but measured in entropic units.

Finally, we mention briefly other interpretations of entropy. As we will see in Chapter 6, models based on the theory of consumers' behaviour assume that consumers maximize utility. If such behaviour is represented by a model which can be generated as an entropy maximizing model, then it is tempting to interpret entropy as utility — and indeed this is sometimes done. Conversely we can argue that since many of the models described in this chapter can be obtained by other means (many of which are described in Chapter 6), the reader can use these alternative theoretical derivations if required without altering the spirit of the mode of application. However, further exploration of this topic will be postponed to Chapter 6.

NOTES ON FURTHER READING

Elementary accounts of the entropy-maximizing method are offered by †Gould (1972), †Haggett, Cliff, and Frey (1977, pp. 40–7) and †Senior (1979). A detailed account is contained in Wilson (1970, †Chapters 1 and 2, *other chapters) and the concept is scrutinized and employed in a variety of ways by *Chapman (1977). A more recent comprehensive treatment is by *Webber (1979). The use of entropy as a measure of diversity is discussed by *Pielou (1969, Chapter 18). Pre-entropy forms of spatial interaction models are reviewed by †Carrothers (1956) and †Olsson (1965) and the post-entropy 'family' of models by †Wilson (1971). †Openshaw (1975) provides a more detailed guide on the use of the models in practice in relation to shopping behaviour.

CHAPTER 5

Disorganized complexity 2: accounting and account-based models

5.1 BASIC CONCEPTS: STATE, CHANGE OF STATE, AND ACCOUNTS

We saw in Section 3.4 how it was often convenient to have labels which characterized the states of system components, say $1, 2, 3, \ldots, p, q, \ldots, N$ (where p and q are used here as typical labels) and then to characterize the system state by a *vector*, say $\{x_p\} = \{x_1, x_2, x_3, \ldots, x_p, x_q, \ldots, x_N\}$ such that x_p was a count of the number of system components in state p.

Accounting is concerned with a particular approach to change of state usually, but not necessarily, over time. An account is drawn up of the numbers of components which change from each state p to each state q in some time period. Where necessary for explicitness, we add time labels: $x_p(t)$ is the number of system components in state p at time t, and $x_p(t + T)$ the number at some time T units later, that is, at time $t + T$. The basic accounting variable is $K_{pq}(t, t + T)$, which is the number of system components in state p at time t *and* in state q at time $t + T$. Thus, for the full range of p and q an array of variables $\{K_{pq}(t, t + T)\}$ is defined which can be written out in full as

$$\{K_{pq}\} = \begin{bmatrix} K_{11} & K_{12} & \ldots & K_{1N} \\ K_{21} & K_{22} & \ldots & K_{2N} \\ \vdots & & & \\ K_{N1} & K_{N2} & \ldots & K_{NN} \end{bmatrix} \tag{5.1}$$

(where the time labels have been dropped, for convenience, but should be 'understood'). This array has the same kind of structure as that for $\{T_{ij}\}$ introduced in the previous chapter. Such arrays are called matrices, and a matrix is sometimes denoted by a single bold letter, as mentioned in Section 3.4. Thus, $\{K_{pq}\}$ can be conveniently designated as $\mathbf{K}$.

As with $\{T_{ij}\}$, the row and column sums of $\mathbf{K}$ have a special significance. Using the asterisk notation introduced in Chapter 4, they can be written

$$p\text{th row sum:} \qquad K_{p*}(t, t + T) = \sum_q K_{pq}(t, t + T) \qquad (5.2)$$

$$q\text{th column sum:} \qquad K_{*q}(t, t + T) = \sum_p K_{pq}(t, t + T). \qquad (5.3)$$

A little thought (or experimentation with a simple example of an array written out in full) shows that the row sum is in fact the total number of system components in state p at time t: that is

$$x_p(t) = K_{p*}(t) = \sum_q K_{pq}(t, t + T) \qquad (5.4)$$

so that the time label on K_{p*} can be changed to t from $(t, t + T)$. Similarly, the qth column sum in the total number of system components in state q at time $t + T$:

$$x_q(t + T) = K_{*q}(t + T) = \sum_p K_{pq}(t, t + T) \qquad (5.5)$$

(and now the time label on K_{*q} is changed to $t + T$). Each row of an accounting array, therefore, does give an *account* of what happens to all the system components originally in the state which labels that row; conversely, each column gives an account of the origin states of all components ending in that particular final state.

To complete the general introduction, it is useful to point out that the state labels may themselves be lists. Thus, if (i, j) described a component state (say home and work locations) then we could take $p = (i, j)$ and $q = (k, l)$ say and define a variable $K_{ijkl}(t, t + T)$, written as K_{pq} for short, as the number of people in home–job state (i, j) at time t and state (k, l) at time $t + T$. Some of the indices may sometimes be arranged as superscripts—as with, say, $K_{jl}^{ik}(t, t + T)$ for the example just given. It is a matter of convenience.

5.2 EXAMPLES OF ACCOUNTING VARIABLES

5.2.1 Population accounts

Consider first the ageing of a population. Define component states as age groups $1, 2, \ldots, r, s, \ldots, R$ (which may be 0–4 years, 5–9, 10–14, etc); then, a basic population accounting array is $\{K_{rs}(t, t + T)\}$, where $K_{rs}(t, t + T)$ is the number of people in age group r at time t and who have aged into group s at time $t + T$. If $P_r(t)$ is the total population in age group r at t, then the row and column sums are, respectively,

$$P_r(t) = K_{r*}(t) = \sum_s K_{rs}(t, t + T) \qquad (5.6)$$

and

$$P_s(t + T) = K_{*s}(t + T) = \sum_r K_{rs}(t, t + T). \qquad (5.7)$$

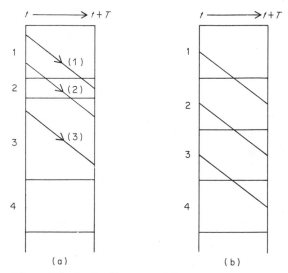

Figure 5.1 Lexis diagrams: (a) uneven age group widths; (b) age group widths equal

It is sometimes useful to represent this kind of ageing in a Lexis diagram, as depicted in Figure 5.1. In such diagrams, both horizontal and vertical axes are calibrated with time — the horizontal one as calendar time and the vertical one with ageing. The divisions on the vertical one represent age groups. A person is represented on this diagram by a lifeline, which flows with time, at 45° to the horizontal. Thus, at a *point* on such a line, the exact age can be read off on the vertical axis which corresponds to the exact calendar time on the horizontal axis. In particular, at time t, the exact age of any person falls into a particular age group; and similarly at time $t + T$. Thus, of the three lifelines shown in Figure 5.1(a), person (1) will be in the $K_{12}(t, t + T)$ group, person (2) in the $K_{13}(t, t + T)$ group, and person (3) in the $K_{33}(t + T)$ group. Note that in this case, $K_{rs}(t, t + T)$ can only be nonzero if $r \leqslant s$: none of us can get any younger.

The case depicted in Figure 5.1(b) is the so-called *simple* case where each age group length is equal to that of the projection period, T(say all five years). It can easily be seen in this case that each surviving person ages into the *next* group, so the only nonzero terms are those of the form $K_{rr+1}(t, t + T)$ and the array takes the special form

$$\mathbf{K} = \begin{bmatrix} 0 & K_{12} & 0 & 0 & \ldots & 0 \\ 0 & 0 & K_{23} & 0 & \ldots & 0 \\ 0 & 0 & 0 & K_{34} & \ldots & 0 \\ \vdots & & & & & \\ 0 & 0 & 0 & & \ldots & K_{RR} \end{bmatrix}$$ (5.8)

(Sometimes the final diagonal term, K_{RR}, is taken as nonzero to close the system.) This representation turns out to be a particularly convenient one for many purposes, but cannot always be used because data do not always come in such a neat form.

The next step in the argument is to introduce space into the accounts. We can assume, as usual a study area made up of discrete zones labelled $1, 2, \ldots, i, j, \ldots, N$. The initial state of a person can then be characterized by age and location, say (r, i) and the final state likewise, say by (s, j). We continue to use subscripts for the age labels and superscripts for the spatial ones, and then the accounting element is $K_{rs}^{ij}(t, t + T)$. The row and column sums, using an obvious notation, are

$$P_r^i(t) = K_{r*}^{i*}(t) = \sum_{js} K_{rs}^{ij}(t, t + T) \tag{5.9}$$

$$P_s^j(t + T) = K_{*s}^{*j}(t + T) = \sum_{ir} K_{rs}^{ij}(t, t + T). \tag{5.10}$$

It is interesting to explore ways of writing out such four-dimensional arrays in full. There are two obvious procedures. The first involves fixing the spatial labels first and then letting the age labels vary within this. The full array can be built up in a neat way if we define *submatrices* for each set of spatial labels as follows:

$$\mathbf{K}^{ij} = \{K_{rs}^{ij}\} = \begin{bmatrix} K_{11}^{ij} & K_{12}^{ij} & \cdots & K_{1R}^{ij} \\ K_{21}^{ij} & K_{22}^{ij} & \cdots & K_{1R}^{ij} \\ \vdots & & & \\ K_{R1}^{ij} & K_{R2}^{ij} & \cdots & K_{RR}^{ij} \end{bmatrix} \tag{5.11}$$

and the full array is then

$$\mathbf{K} = \{\mathbf{K}^{ij}\} = \begin{bmatrix} \mathbf{K}^{11} & \mathbf{K}^{12} & \cdots & \mathbf{K}^{1N} \\ \mathbf{K}^{21} & \mathbf{K}^{22} & \cdots & \mathbf{K}^{2N} \\ \vdots & & & \\ \mathbf{K}^{N1} & \mathbf{K}^{N2} & \cdots & \mathbf{K}^{NN} \end{bmatrix} \tag{5.12}$$

where each 'element' is itself a (sub)matrix of the form (5.11). The full array can be written out explicitly as in (5.13) below
If this large array seems complicated, the reader should take a specific example — say $R = 3$, $N = 2$, and write out the array in full for that example.

$$
\mathbf{K} = \begin{bmatrix}
K_{11}^{11} & K_{12}^{11} \dots K_{1R}^{11} & K_{11}^{12} & K_{12}^{12} \dots & & K_{11}^{1N} & K_{12}^{1N} \dots \\[2mm]
K_{21}^{11} & K_{22}^{11} \dots K_{2R}^{11} & K_{21}^{12} & K_{22}^{12} \dots & \dots & K_{21}^{1N} & K_{22}^{1N} \dots \\[2mm]
K_{R1}^{11} & K_{R2}^{11} \dots K_{RR}^{11} & \vdots & \vdots & & \vdots & \vdots \\[3mm]
K_{11}^{21} & K_{12}^{22} \dots & K_{11}^{22} & K_{12}^{22} \dots & & K_{11}^{2N} & K_{12}^{2N} \dots \\[2mm]
K_{21}^{21} & K_{22}^{21} \dots & K_{21}^{22} & K_{22}^{22} \dots & & K_{21}^{2N} & K_{22}^{2N} \dots \\[2mm]
\vdots & \vdots & \vdots & \vdots & & \vdots & \vdots \\[3mm]
& \vdots & & \vdots & & & \vdots \\[3mm]
K_{11}^{N1} & K_{12}^{N1} \dots & K_{11}^{N2} & K_{12}^{N2} \dots & & K_{11}^{NN} & K_{12}^{NN} \dots \\[2mm]
K_{21}^{N1} & K_{22}^{N1} \dots & K_{21}^{N2} & K_{22}^{N2} \dots & & K_{21}^{NN} & K_{22}^{NN} \dots \\[2mm]
\vdots & \vdots & \vdots & \vdots & & \vdots & \vdots
\end{bmatrix}
\tag{5.13}
$$

The second method involves fixing the age labels first. This would involve defining submatrices of the form

$$
\mathbf{K}_{rs} = \begin{bmatrix}
K_{rs}^{11} & K_{rs}^{12} \dots K_{rs}^{1N} \\
\vdots & \\
K_{rs}^{N1} & K_{rs}^{N2} \dots K_{rs}^{NN}
\end{bmatrix}
\tag{5.14}
$$

and the full array as

$$
\mathbf{K} = \begin{bmatrix}
\mathbf{K}_{11} & \mathbf{K}_{12} \dots \mathbf{K}_{1R} \\
\mathbf{K}_{21} & \mathbf{K}_{22} \dots \mathbf{K}_{2R} \\
\vdots & \\
\mathbf{K}_{R1} & \mathbf{K}_{R2} \dots \mathbf{K}_{RR}
\end{bmatrix}
\tag{5.15}
$$

and this could also be written out in full in the same manner as (5.13) from (5.12). The reader can also easily check that the row and column sums remain the same whichever of the two representations is used.

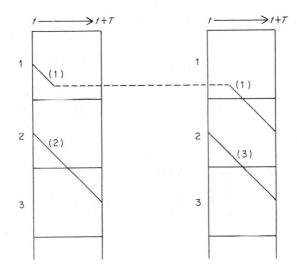

Figure 5.2 Lexis diagram, with migration

It is also useful to note that the Lexis diagram can be extended to the multiregion case as indicated for two regions in Figure 5.2. A migration then involves a lifeline such as (1) which is depicted as a shift from one diagram to another. Such a person would appear in the $K_{12}^{12}(t, t + T)$ accounting element.

The perceptive reader will have noticed that the population accounts presented so far are incomplete: there is no mention of births and deaths; and of course these are events which occur continuously *during* the period rather than at the end points alone. This problem is dealt with by introducing new initial states, called birth states and labelled $\beta(i), i = 1, 2, \ldots, N$ (which stands for 'birth in region i during the period'), and new final states, death states, labelled $\delta(j), j = 1, 2, \ldots, N$ (which stands for 'death in region j during the period'). There are four types of combinations of the states which have now been defined: K^{ij} (existence at t and $t + T$, possible migration if $i \neq j$); $K^{i\delta(j)}$ (existence at t, death at j before $t + T$, possibly preceded by migration if $j \neq i$); $K^{\beta(i)j}$ (birth in i during the period, existence at $t + T$, migration if $i \neq j$); $K^{\beta(i)\delta(j)}$ (birth followed by death during the period, with an intervening migration if $i \neq j$—essentially the infant mortality problem).

Age labels can be introduced provided they are defined carefully for the new states $\{K_{rs}^{ij}\}$ is exactly as before; in $\{K_{rs}^{i\delta(j)}\}$, s is taken as age at death; in $\{K_{rs}^{\beta(i)j}\}$, r is in a sense redundant, since age at birth is zero, but it turns out to be convenient to use this for age of mother at the birth; then, in $\{K_{rs}^{\beta(i)\delta(j)}\}$, r is the age of the mother at the birth and s is the age of the child at death. With these definitions, we can then take $\mathbf{K}^{ij}$, $\mathbf{K}^{i\delta(j)}$, $\mathbf{K}^{\beta(i)j}$, and $\mathbf{K}^{\beta(i)\delta(j)}$ to be submatrices. The submatrices have the following general structures: $\mathbf{K}^{ij}$ as in (5.11), but we now note that the subdiagonal elements *must be zero* (or people could get younger):

$$\mathbf{K}^{ij} = \begin{bmatrix} K^{ij}_{11} & K^{ij}_{12} \ldots & K^{ij}_{1R} \\ 0 & K^{ij}_{12} \ldots & K^{ij}_{2R} \\ \vdots & & \\ 0 & 0 & \ldots & K^{ij}_{RR} \end{bmatrix} \tag{5.16}$$

$\mathbf{K}^{i\delta(j)}$ is similar:

$$\mathbf{K}^{i\delta(j)} = \begin{bmatrix} K^{i\delta(j)}_{11} & K^{i\delta(j)}_{12} \ldots K^{i\delta(j)}_{1R} \\ 0 & K^{i\delta(j)}_{22} \ldots K^{i\delta(j)}_{2R} \\ \vdots & & \\ 0 & 0 & \ldots K^{i\delta(j)}_{RR} \end{bmatrix} \tag{5.17}$$

The birth matrices have a distinctive shape because of the first age label being associated with the mother:

$$\mathbf{K}^{\beta(i)j} = \begin{bmatrix} 0 & 0 & \ldots 0 \\ \vdots & & \\ K^{\beta(i)j}_{\mu 1} & K^{\beta(i)j}_{\mu 2} \ldots \\ \vdots & & \\ K^{\beta(i)j}_{\nu 1} & K^{\beta(i)j}_{\nu 2} \ldots \\ 0 & 0 & \ldots \\ \vdots & & \\ 0 & 0 & \ldots 0 \end{bmatrix} \tag{5.18}$$

where μ and ν are the lower and upper limits of the child-bearing age groups. Similarly

$$\mathbf{K}^{\beta(i)\delta(j)} = \begin{bmatrix} 0 & 0 & \ldots 0 \\ \vdots & & \\ K^{\beta(i)\delta(j)}_{\mu 1} & K^{\beta(i)\delta(j)}_{\mu 2} \ldots \\ \vdots & & \\ K^{\beta(i)\delta(j)}_{\nu 1} & K^{\beta(i)\delta(j)}_{\nu 2} \ldots \\ 0 & 0 & \ldots 0 \\ \vdots & & \\ 0 & 0 & \ldots 0 \end{bmatrix} \tag{5.19}$$

If we denote the corresponding arrays by $\{K^{ij}\}$, $\{K^{i\delta(j)}\}$, $\{K^{\beta(i)j}\}$, and $\{K^{\beta(i)\delta(j)}\}$ the full set of accounts takes the form

$$K = \begin{bmatrix} \{K^{ij}\} & \{K^{i\delta(j)}\} \\ \hline \{K^{\beta(i)j}\} & \{K^{\beta(i)\delta(j)}\} \end{bmatrix} \tag{5.20}$$

To write the array out in full without using the submatrix notation needs a large sheet of paper. However, it may be a useful exercise for the reader to do this, say for $N = 2$, $R = 4$, $\mu = 2$, $v = 3$.

The great advantage of this systematic approach to the development of an accounting system is that nothing is neglected and that the principles of account building and the way the notation is developed guarantee this. It would be very easy without this to forget to deal with something like $K^{\beta(i)\delta(j)}, i \neq j$ flows — nonsurviving infants who have migrated before dying. The significance of all this will be more evident in the section on population modelling below.

5.2.2 Economic accounts

Economies are concerned with the flows of goods (with 'good' to be interpreted as widely as possible — to include, for example, the production and consumption of services). The economy is divided into sectors $1, 2, \ldots, m, n, \ldots, M$ say. Then, the nonspatial accounting variable is Z^{mn}, the flow of goods from sector m to sector n. Some common units are assumed, such as money-value or employment per unit product. This accounts for all 'intermediate demand' — the goods used as inputs by other sectors, and Y^m is usually defined to be 'final' demand (by consumers) for the product of sector m. The total product is defined as X^m. Then, the accounts can be written as

$$Z = \begin{bmatrix} Z^{11} & Z^{12} & \ldots & Z^{1m} \\ Z^{21} & Z^{22} & \ldots & Z^{2m} \\ \vdots & & & \\ Z^{m1} & Z^{m2} & \ldots & Z^{mm} \end{bmatrix}, \quad Y = \begin{bmatrix} Y^1 \\ Y^2 \\ \vdots \\ Y^m \end{bmatrix}, \quad X = \begin{bmatrix} X^1 \\ X^2 \\ \vdots \\ X^m \end{bmatrix} \tag{5.21}$$

where *column vectors* have been defined for final demand and total product. The row accounting equation is not now simply the row sum of Z, since final demand has been separately identified. Clearly

$$\sum_n Z^{mn} + Y^m = X^m \tag{5.22}$$

is the appropriate equation.

The next step, as in the population case, is to add a spatial dimension. Y_i^m and X_i^m have obvious definitions as final demand for m and total product in zone i. Z_{ij}^{mn} is the amount of the product of m in zone i used by sector n in zone j. The array could be written out in full by defining submatrices Z_{ij} whose (m, n)th element was Z_{ij}^{mn}. The row accounting relationship would be

$$\sum_{jn} Z_{ij}^{mn} + Y_i^m = X_i^m. \tag{5.23}$$

The approach adopted so far involves, implicitly, a number of simplifying assumptions. In equation (5.23) for example, it is assumed that none of the product of sector m in i goes directly to *final demand* in other regions. This may be reasonable provided sectors are carefully defined and 'exports' of this kind go via an agency. There is another more serious problem. Either we are assuming that each sector has a unique product, labelled with the same name as the sector, or that the various products of a sector can be lumped together in a bundle, valued in money terms, and thereafter treated as a good. It is better to be more explicit about this and to define an array such as $\{x_{ij}^{mnk}\}$ — the flow of commodity (or good k from sector m in i to sector n in $j(k = 1, 2, ..., K$ — other limits as before). If sectors are broadly defined, to include one or more for people as well as for organizations, then a complete set of accounts is implicit in this array.

To explore this set of accounts further, let us first drop the spatial labels and consider a single region: $\{x^{mnk}\}$ is the array. To write this out in full would need $K\,M \times M$ matrices. It is very rare, however, that data are available at this level of detail and it is customary to define two summary matrices, the so-called *absorption* matrix, $\{U^{kn}\}$ and the *make* matrix, $\{V^{mk}\}$, whose typical elements are defined by

$$U^{kn} = \sum_m x^{mnk} \tag{5.24}$$

representing the absorption of k by n from all sources and

$$V^{mk} = \sum_n x^{mnk}$$

representing the amount of k produced in sector m. These two matrices can be put together in a dog-leg array as shown in Figure 5.3. As we shall see below, so-called rectangular input output models can be constructed on the basis of this array.

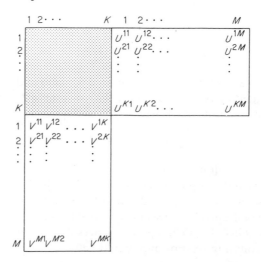

Figure 5.3 A rectangular input–output table

The spatial labels could then be added again and the accounts set out using submatrices in the usual way.

5.2.3 Ecological accounts

We have seen that accounts are concerned with tracing entities through different states — literally 'accounting' for them at the beginning and end of a time period. It turns out that one of the most popular forms of ecological models can be seen as based on accounting concepts: the so-called 'compartment' models.

A compartment is a part of the ecosystem. The whole system is broken down, exclusively, into a number of compartments, together with its environment. For example, the ecosystem represented in Figure 1.11 could be considered to consist four compartments, as depicted in Figure 5.4. The boxes have been numbered 1–4 and the main flows are shown, including the main flows to and from the environment (marked as 'E').

In a compartment model, flows are followed between compartments. There can be flows of energy or nutrients, for example. Each compartment, therefore, can be considered as a state in an accounting sense. For a flow k, therefore, we could define x_{mn}^k as the flow from compartment m to compartment n during a time period; or, to put it in the more usual way, as a statement of the quantity of k in compartment m at time t which was in compartment n at time $t + \delta t$, say. The environment can be represented by the symbol e and flows to and from it by x_{me}^k and x_{en}^k respectively.

If we now let $S_m^k(t)$ be the total stock of k in m at time t, then

$$S_m^k(t) = \sum_n x_{mn}^k + x_{me}^k \tag{5.26}$$

and

$$S_n^k(t + \delta t) = \sum_m x_{mn}^k + x_{en}^k \tag{5.27}$$

are the usual accounting totals at the beginning and end of the time period. The

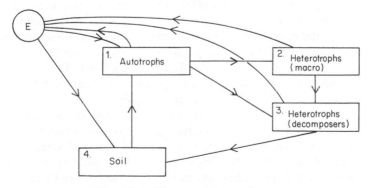

Figure 5.4 Elements of an ecosystem

increment in stock of k is obtained by subtraction:

$$S_n^k(t + \delta t) - S_n^k(t) = \sum_m x_{mn}^k - \sum_m x_{nm}^k + x_{en}^k - x_{ne}^k \tag{5.28}$$

(where we have interchanged m and n in equation (5.26) to get $S_n^k(t)$ rather than $S_m^k(t)$).

Suppose, for example, we return to Figure 5.4 and consider nitrogen flows (labelled as $k = N$). The accounting equations (5.28), for each compartment, are

$$S_1^N(t + \delta t) - S_1^N(t) = x_{e1}^N - x_{12}^N - x_{13}^N \tag{5.29}$$

$$S_2^N(t + \delta t) - S_2^N(t) = x_{12}^N - x_{23}^N \tag{5.30}$$

$$S_3^N(t + \delta t) - S_3^N(t) = s_{13}^N - x_{23}^N - x_{34}^N \tag{5.31}$$

$$S_4^N(t + \delta t) - S_4^N(t) = x_{e4}^N + x_{34}^N - x_{41}^N . \tag{5.32}$$

We will consider this set of equations further in exploring account-based ecological modelling in Section 5.4.5.

5.2.4 Accounts in water resource models

If a river system is sectionalized into reaches, then clearly there are conservation relations which could be set up as accounting equations. However, it is more interesting to establish a link with the previous subsection and to consider a compartment model within which reaches are distinguished. Here, we follow Kelly and Spofford (1977) and this also illustrates an alternative approach to compartment modelling. The addition of ecosystem compartments to a water model has the advantage that we can extend our resources modelling range to cover pollution because we can represent directly the biological effects of pollutants being added to the river — e.g. as domestic or industrial waste.

In the previous subsection, compartments were biological entities and we sought to model the flow of nutrients, or energy, or other basic constituents, between compartments. In the approach of Kelly and Spofford, the compartments again represent basic biological compartments, this time identifying separate compartments for the main nutrients. The 'stock' in each compartment is measured by the biomass. Thus, the accounting relationships are concerned with conservation of mass and the modelling problem with the direct representation of biological processes which affect biomass. We outline the accounting system here and the associated modelling problems in Section 5.4.7.

For each reach, eleven compartments are defined: the main eight are made up of algae (primary producers, fixing energy from photosynthesis), zooplankton (herbivores), fish (carnivores), and bacteria (decomposers) together with compartments for nitrogen, phosphorus, oxygen, and organic matter. The main biological reactions take place among these eight types. The remaining three represent turbidity, toxicity, and heat. These are not affected themselves by the biological processes, but their levels affect the process-rates in other compartments.

The animal or plant equations take the form

$$x_{mt+\delta t} - x_{mt} = f_m - r_m - d_m - e_m - p_m + y_m - wx_m \qquad (5.33)$$

for $m = 1-4$ say (though with the last three terms omitted for the fish equation). f_m is the feeding rate (though it is actually a volume rather than a rate) for m, r_n the respiration rate, d_m the death rate, e_m the excretion rate, p_m the predation rate, y_m the input from outside, and w the water throughflow rate.

The oxygen equation depends on the production rate from algae and the use in respiration. Through bacteria, it helps break down organic matter, and also, through chemical processes, certain kinds of toxic material. It is this which leads to the measure of organic material in terms of B.O.D.—biological oxygen demand.

Organic matter increases through death, excretion, and exogenous inputs. Nitrogen and phosphorus are produced in amounts proportional to respiration rates, less use by algae in photosynthesis, and chemical reactions involving toxic materials.

To take this model further, the rates on the right-hand side of equations like (5.33) need to be modelled, and we pursue this question in Section 5.4.7.

5.3 SOME GENERAL NOTES ON ACCOUNTING VARIABLES

Accounting variables are intimately linked to the notion of component state description and to levels of resolution. Such state indices have to be defined, as we saw in Chapter 2, in spatial, sectoral, and temporal dimensions. In the preceding section we saw examples of all three seeking to represent component change of sector or location within a time period. When more than one index is involved in the state description, we saw in Section 5.1 that this can if necessary be given the single label p which then stands for an index list. One advantage of this notion is that it enables us to see clearly which indices are intrinsically part of the accounts and which are merely labels. Those which are part of p or q in a K_{pq} formulation are the former; otherwise, the latter. For example, with K_{rs}^{ij}, $p = (i, r)$, $q = (j, s)$ and there are no additional labels. *But with* x_{ij}^{mnk}, $p = (i, m)$, $q = (j, n)$ and k is a label defining a set of accounts for each commodity. With the transport accounts T_{ij}^{kn}, which are essentially purely spatial accounts, $p = i$, $q = j$ and both k and n are labels (for mode and person type).

5.4 SIMPLE ACCOUNT-BASED MODELS

5.4.1 Principles for model building

Suppose we have a set of accounts $\{K_{pq}(t, t + T)\}$ for suitably defined states. Then we can define row-sum rates by taking any element and dividing by the corresponding row sum. Let the array of such rates be denoted by $\{R_{pq}\}$ (with $(t, t + T)$ implicit henceforth) with

$$R_{pq} = K_{pq}/K_{p*}. \qquad (5.34)$$

This is the proportion of system components in state p at time t who are in state q at time $t + T$. It can also be interpreted as the *probability of the transition from state p to state q* during the period t to $t + T$. Define

$$G_{pq} = R_{qp}. \qquad (5.35)$$

Then, the following equation is an identity:

$$K_{*q} = \sum_p G_{qp} K_{p*} \qquad (5.36)$$

since, using equations (5.35) and (5.34),

$$\sum_p G_{qp} K_{p*} = \sum_p R_{pq} K_{p*}$$

$$= \sum_p \frac{K_{pq}}{K_{p*}} K_{p*}$$

$$= \sum_p K_{pq}$$

$$= K_{*q}. \qquad (5.37)$$

Thus, an identity such as (5.36) can be constructed, with rates $\{G_{pq}\}$ obtained from the accounts, which relates time t distributions, $K_{p*}(t)$, to time $t + T$ distributions $K_{*q}(t + T)$. Such an identity becomes a *model* if some independent rule for obtaining G_{pq}'s can be established. The importance of the accounting base of such a model is that, first, the *structure* of the array of rates is correct — all the p–q transitions are properly identified; and, secondly, the rates can be properly *measured* (by using equations (5.34) and (5.35)) for any period for which accounting data is available.

The simplest *rule*, of course, is to assume that the rates remain constant over time. Such a model is the simplest form of Markov model and it turns out to have an important property: that after a sufficiently long period (t to $t + nT$ for suitably large n, say), the number of system components in different states remains constant. It is also perhaps significant that Markov was working in the late nineteenth and early twentieth centuries at the same time as those who were laying the foundations of statistical mechanics because this modelling method is, in effect, another kind of statistical averaging method which relies on Weaver-II disorganized complexity methods to hold for its validity. However, the utility of the accounting base for modelling extends beyond systems for which such assumptions are true, as we shall see in the next section.

The methods of account-based modelling can be stated more succinctly if matrix algebra is used. Just occasionally, even in the elementary treatment of this chapter, it is essential, but the reader should be able to cope with most of the chapter without it. For those readers who have the tool available, however, the main results will be stated in matrix form. For example, taking $\mathbf{K} = \{K_{pq}\}$,

$\mathbf{R} = \{R_{pq}\}$, $\mathbf{G} = \{G_{pq}\}$, equations (5.35) and (5.36) can be rewritten as

$$\mathbf{G} = \mathbf{R}' \tag{5.38}$$

(where the prime denotes transposition) and

$$\mathbf{P}(t + T) = \mathbf{G}\mathbf{P}(t) \tag{5.39}$$

where $\mathbf{P}(t)$ is the vector $\{P_p(t)\}$ and

$$P_p(t) = K_{p*}(t) \tag{5.40}$$

$$P_p(t + T) = K_{p*}(t + T) = K_{*q}(t). \tag{5.41}$$

For obvious reasons, $\mathbf{G}$ in equation (5.39) is sometimes known as the *growth* or *change* matrix.

5.4.2 Example 1: social mobility

A large number of simple, but useful, examples can be developed directly from the argument of the previous section. Suppose states $1, 2, \ldots, p, q, \ldots, Q$ represent social classes and $\{K_{pq}\}$ is an array of transition data obtained from a social survey. Then, the model presented above (in equations (5.36) and (5.39)) applies directly. If the total population is assumed fixed, then it may also be stated in terms of proportions in a class at a time, say using

$$p_p(t) = P_p(t)/P_*(t) \tag{5.42}$$

so that the model equation would become

$$\mathbf{p}(t + T) = \mathbf{G}\mathbf{p}(t). \tag{5.43}$$

Bartholomew (1967) works in this way and gives numerical examples for this case. Many other examples are more complicated in some way, however, and it is to these we now turn.

5.4.3 Example 2: the Rogers' model of population growth

Rogers (1966, 1971, 1975), building on the work of Leslie (1945), constructed a model of a multiregional population system. Let $P_r^i(t)$ be the population in zone i, age group r, at time t. Assume that age group intervals are each equal to the projection period T so that all ageing is into the next highest age group. Then, the model takes the form

$$P_1^i(t + T) = \sum_j \sum_{r=\mu}^{v} b_{r1}^{ij} P_r^j(t) \tag{5.44}$$

$$P_r^i(t + T) = S_{rr-1}^i P_{r-1}^i(t) + \sum_{j \neq i} m_{rr-1}^{ij} P_{r-1}^j(t), \ r > 1. \tag{5.45}$$

This can be written out in matrix form as

$$
\begin{bmatrix} P_1^1(t+T) \\ P_2^1(t+T) \\ \vdots \\ P_R^1(t+T) \\ P_1^2(t+T) \\ \vdots \\ P_R^2(t+T) \\ \\ \vdots \\ \\ P_1^N(t+T) \\ \vdots \\ P_R^N(t+T) \end{bmatrix}
=
\left[\begin{array}{ccc|ccc|c|c}
0 & 0 \ldots b_{\mu1}^{11} \ldots b_{v1}^{11} \ldots 0 & 0 & 0 \ldots b_{\mu1}^{12} \ldots b_{v1}^{12} \ldots 0 & & \\
S_{21}^1 & 0 \ldots 0 & & m_{21}^{12} & & \\
0 & S_{32}^1 & & & m_{32}^{12} & & \\
& & S_{RR-1}^1 \; 0 & & & & \\
\hline
m_{21}^{21} & & & & & \\
& m_{32}^{21} & & & & \\
& & & & & \\
\hline
& & & & & \\
\hline
& & & & & \\
\end{array}\right]
\begin{bmatrix} P_1^1(t) \\ P_2^1(t) \\ \vdots \\ P_R^1(t) \\ P_1^2(t) \\ P_2^2(t) \\ P_R^2(t). \\ \\ \\ \\ P_1^N(t) \\ P_2^N(t) \\ P_R^N(t) \end{bmatrix}
$$

This model clearly takes the form

$$P(t+T) = GP(t) \tag{5.47}$$

for suitably defined P and G. The rates within the model have obvious interpretations as birth rates (b_r^{ij}), migration rates (m_{rr-1}^{ij}) or survival rates (s_{rr-1}^{ij}). Note in these definitions that the state changes (for location and age) read from right to left instead of the more usual left to right. This is so that they can be located in the 'correct' position to the G-matrix in equation (5.46) but arises because of the need to transpose the matrix of account-based rates to obtain G as we saw in Section 5.4.1. This now raises the question: what is the accounting basis of this model and what use would it be?

The answer to this question lies in the accounting array (5.20). Rates R can be obtained from this array by dividing by the row sums in the usual way. However, for the lower half, such row sums would be total births, and this is not the usual, or a helpful, birthrate, and so total births, $K^{\beta(i)*}$ terms, is replaced by total population terms, K^{i*}. The matrix of rates thus obtained takes the form

$$
R = \left[\begin{array}{c|c}
\left\{ K_{rs}^{ij}/K_{r*}^{i*} \right\} & \left\{ K_{rs}^{i\delta(j)}/K_{r*}^{i*} \right\} \\
\hline
\left\{ K_{r1}^{\beta(i)j}/K_{r*}^{i*} \right\} & \left\{ K_{rs}^{\beta(i)\delta(j)}/K_{r*}^{i*} \right\}
\end{array}\right]
\tag{5.48}
$$

and **G** can be obtained by transposition. If an explicit account is *not* needed of deaths, the right-hand half of **R** can be ignored for the purposes of constructing the model (but will play another role which will be seen shortly), and since the birth rate terms multiply into total population terms, the remaining two quarters of the rates matrix can be combined (by adding the ith and $(N + i)$th row for each i). This **G** then reduces to **G** in (5.46)

$$S^i_{rr-1} = K^{ii}_{r-1r}/K^{i*}_{r-1*} \tag{5.49}$$

$$m^{ij}_{rr-1} = K^{ji}_{r-1r}/K^{j*}_{r-1*} \tag{5.50}$$

and

$$b^{ij}_{r1} = K^{\beta(j)i}_{r1}/K^{j*}_{r*}. \tag{5.51}$$

This rather complicated procedure, which has been described only briefly here, is described more fully elsewhere (e.g. Wilson, 1974, Chapter 7, where the reader is taken through it step by step). However, the two uses of an accounting base for a model can now be seen: the rates in the model are properly identified in terms of accounting elements, as in equations (5.49)–(5.51), and this helps with measurement. Both of these features are important in this case. In some early versions of the Rogers model, for example, migrating infant terms, arising from $b^{ij}, i \neq j$ rates, were neglected. And, for example, K^{ii} is not normally available directly from data, but it can now be obtained by using the accounting relationship

$$K^{ii}_{rs} = K^{i*}_{r*} - \sum_{j \neq i} K^{ij}_{rs} - \sum_j K^{i\delta(j)}_{rs} \tag{5.52}$$

if all the quantities on the right-hand side are known. (This causes difficulties in practice which are taken up in Section 5.5.) Note that this is *not* one of the obvious definitions of region i survivors, which would be

Survivors (K^{ii}_{rs}) = Original population (K^{i*}_{r*}) – out migrants $\left(\sum_{j \neq i} K^{ij}_{rs}\right)$

because

$$- \text{ total deaths in } i(K^{*\delta(i)}_{*s}) \tag{5.53}$$

$$K^{*\delta(i)}_{*s} \neq \sum_j K^{i\delta(j)}_{rs} \tag{5.54}$$

(the right-hand side being the quantity needed in equation (5.52)). Thus, it is very easy for the wrong quantity to be used in the measure of the survival rate in equation (5.49) unless the accounting foundation is used.

5.4.4 Example 3: the input–output model

The simplest input–output model is based on the accounts and vectors defined at (5.21) and the accounting equation (5.22). The rate definition used here is

$$a^{mn} = Z^{mn}/X^n \tag{5.55}$$

which is a column-based rate if common units are used throughout the accounts and $\mathbf{Z}$ includes primary inputs*. a^{mn} is the amount of good m needed to produce a unit of good n — that is, needed as an input to the process which produces n. A row sum rate would have been the proportion of the total product of a good m which was used to produce n. For an economy, it seems better to rely on the column rate and to 'suck in' inputs to a production process, especially when 'constant rate' (or 'coefficient' as these rates are more often called) assumptions are used. $\{a^{mn}\}$ is assumed to reflect the technical structure of the economy.

A model equation can now be formed from the accounting equation (5.22) as follows. Equation (5.55) can be rearranged to give

$$Z^{mn} = a^{mn} X^n. \tag{5.56}$$

Substitute for Z^{mn} in (5.22)

$$\sum_n a^{mn} X^n + Y^m = X^m. \tag{5.57}$$

This can be rearranged as

$$X^m - \sum_n a^{mn} X^n = Y^m \tag{5.58}$$

and this is a set of M linear simultaneous equations in $\{X^m\}$ which can be solved for given $\{a^{mn}\}$ and $\{Y^m\}$. Thus, $\{a^{mn}\}$ can be measured using accounts obtained in a survey and either assumed constant or adjusted slightly for any known changes (see Section 5.5.2). Then, for any given Y^m, the model equation (5.58) can be solved for the total product in each sector X^m.

The argument can be taken further with the use of some matrix algebra. First, it is useful to define the Kronecker delta, δ_{mn}:

$$\delta_{mn} = \begin{cases} 1 & \text{if } m = n \\ 0 & \text{otherwise.} \end{cases} \tag{5.59}$$

Equation (5.58) can then be written

$$\sum_n (\delta_{mn} - a^{mn}) X^n = Y^m. \tag{5.60}$$

The unit matrix $\mathbf{I}$ is that with ones in the diagonal and zeros elsewhere — in fact with δ_{mn} as its (m, n)th element. $\mathbf{A}$ can be taken as the matrix whose elements are $\{a_{mn}\}$. Then (5.60) can be written in matrix form as

$$(\mathbf{I} - \mathbf{A})\mathbf{X} = \mathbf{Y} \tag{5.61}$$

and this can be solved by inverting the matrix $(\mathbf{I} - \mathbf{A})$ to give

$$\mathbf{X} = (\mathbf{I} - \mathbf{A})^{-1} \mathbf{Y}. \tag{5.62}$$

Further insight can be obtained if we recall that a matrix of the form $(\mathbf{I} - \mathbf{A})^{-1}$

---

*Such a model will work if X^n is as defined in equation (5.22) and the elements of each *row* in the accounts are in the same units, though the units of each a^{mn} have then to be carefully defined.

can be expanded:

$$(I - A)^{-1} = I + A + A^2 + A^3 + \ldots \tag{5.63}$$

so that, substituting in (5.62),

$$X = Y + AY + A^2 Y + A^3 Y + \ldots \tag{5.64}$$

Thus, the total product X is final demand, plus intermediate demand needed to produce Y, the next term AY, plus $A(AY)$—which is needed to produce AY, and so ad infinitum. Hence, $(I - A)^{-1}$ is known as the *multiplier*.

A *rectangular* input–output model can be obtained from the set of accounts presented in Figure 5.3, based on the absorption and make matrices $\{U^{kn}\}$ and $\{V^{mk}\}$. Let Y^k be the final demand for good k and Q^k and X^m the total amounts produced by good and sector respectively. Rates can be constructed from absorption and make matrices as follows:

$$b^{kn} = U^{kn}/X^n \tag{5.65}$$

which gives

$$U^{kn} = b^{kn} X^n \tag{5.66}$$

for later use, and

$$d^{mk} = V^{mk}/Q^k \tag{5.67}$$

which gives

$$V^{mk} = d^{mk} Q^k. \tag{5.68}$$

The accounting equations take the form, as usual,

$$\text{Intermediate demand} + \text{final demand} = \text{total product} \tag{5.69}$$

and can be stated separately by good and sector. Final demand by sector m is $\sum_k d^{mk} Y^k$, using d^{mk} from (5.67). Intermediate demand by good for sector m can be taken as b^{km} times the total amount of activity in m (which is taken as $\sum_l V^{ml}$) $- b^{km} \sum_l V^{ml}$, and hence $\sum_{ml} b^{km} V^{ml}$ in all. Similarly $\sum_{kn} d^{mk} U^{kn}$ is the intermediate demand by sector m. Thus the two sets of accounting equations are

$$\sum_{ml} b^{km} V^{ml} + Y^k = Q^k \tag{5.70}$$

$$\sum_{kn} d^{mk} U^{kn} + \sum_k d^{mk} Y^k = X^m. \tag{5.71}$$

Substituting for V^{ml} from (5.68) into (5.70) and for U^{kn} from (5.66) into (5.71), we get

$$\sum_{ml} b^{km} d^{ml} Q^l + Y^k = Q^k \tag{5.72}$$

$$\sum_{kn} d^{mk} b^{kn} X^n + \sum_k d^{mk} Y^k = X^m. \tag{5.73}$$

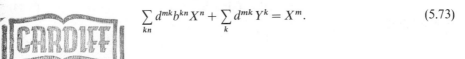

Using Kronecker deltas, these equations can be written

$$\sum_l (\delta^{kl} - \sum_m b^{km} d^{ml}) Q^l = Y^k \tag{5.74}$$

$$\sum_n (\delta^{mn} - \sum_k d^{mk} b^{kn}) X^n = \sum_k d^{mk} Y^k \tag{5.75}$$

which are two sets of linear simultaneous equations in $\{Q^l\}$ and $\{X^n\}$ respectively. In matrix form, these can be written

$$(\mathbf{I} - \mathbf{bd})\mathbf{Q} = \mathbf{Y} \tag{5.76}$$

$$(\mathbf{I} - \mathbf{db})\mathbf{X} = \mathbf{dY} \tag{5.77}$$

and solved to give

$$\mathbf{Q} = (\mathbf{I} - \mathbf{bd})^{-1}\mathbf{Y} \tag{5.78}$$

$$\mathbf{X} = (\mathbf{I} - \mathbf{db})^{-1}\mathbf{dY}. \tag{5.79}$$

Note that the 'rectangularness' of the arrays $\mathbf{U}$ and $\mathbf{V}$ and associated matrices of coefficients $\mathbf{b}$ and $\mathbf{d}$ is removed in the equations because the coefficients turn up as products $\mathbf{bd}$ and $\mathbf{db}$, which are square. (Only square matrices can be inverted of course). Thus, $(\mathbf{I} - \mathbf{bd})^{-1}$ and $(\mathbf{I} - \mathbf{db})^{-1}$ are the matrix multipliers arising from this new set of accounts.

5.4.5 Comparison of simple model types

We have so far identified two distinct types of simple account-based model which can be summarized in equation form as

$$\mathbf{P}(t + T) = \mathbf{GP}(t) \tag{5.80}$$

$$\mathbf{X} = (\mathbf{I} - \mathbf{a})^{-1}\mathbf{Y}. \tag{5.81}$$

The distinguishing feature is not so much that one is based on row rates and the other on column rates but that in the second case, an element of the accounts, the final demand, $\mathbf{Y}$, is distinguished from the other transactions. This can be seen more clearly if we manipulate the population model as follows. Remove the elements from $\mathbf{G}$ which are concerned with birth and in-migration and call the matrix without these elements $\mathbf{C}$. Let the births and in-migrants be part of a separate vector, $\mathbf{b}$. Then equation (5.80) can be written

$$\mathbf{P}(t + T) = \mathbf{CP}(t) + \mathbf{b}. \tag{5.82}$$

In the long run, if a steady state is reached, we would have

$$\mathbf{P}(t + T) = \mathbf{P}(t) \tag{5.83}$$

so that *in this case*

$$\mathbf{P} = \mathbf{CP} + \mathbf{b}. \tag{5.84}$$

Rearranging,

$$(\mathbf{I} - \mathbf{C})\mathbf{P} = \mathbf{b} \tag{5.85}$$

and so

$$P = (I - C)^{-1}b. \tag{5.86}$$

A model of the first type has been transformed into a model of the second by distinguishing the births and in-migrants, b.

In the population case, if all the coefficients are assumed constant, then there is an equilibrium population P such that

$$P = GP. \tag{5.87}$$

(Technically, P is the eigenvector of G whose eigenvalue is 1, and Markov theory shows this to be the case.) In such a case, the model could be written either in form (5.80) or in form (5.86) provided b was taken as the vector of births and in-migrants implied by (5.80). This suggests that the issue in model design is not the case where all coefficients are constant — when the two forms can be made equivalent — but the case where they are not constant. In the input–output model, for example, constant rates can be assumed for the technological structure, but to incorporate final demand into the transactions' accounts on the same basis — with final demand as a fixed proportion of total output for each sector — would be a nonsense. In this case, the model is *driven by* the exogenous final demand vector. In the population case, however, there is not usually the same argument for assuming that all the rates *except* those associated with births and in-migration are constant. There may be special instances where it does apply, as in Stone's (1970) work on university populations which produced the model given by equation (5.86).

5.4.6 Example 4: rates for ecological 'compartment' models

In the case of population and economic accounts, it was possible to obtain suitable rates by dividing by the appropriate row or column sums and to use these rates as the basis for predictive models. In the case of birth and death rates in population models, the row sum provides the appropriate denominator because it is the original population which is the generator of the events. In the economic case, the column sum is correct because it is the receiving sector (taking a good as an input) which determines how much is required. These two mechanisms also occur in ecological modelling, where they are known as donor-controlled and recipient-controlled respectively. Patten (1971) has usefully distinguished eight different kinds of flow-mechanism (and, of course, other more complicated ones could be devised).

They are:

(i) $F_{mn} = k$: the flow from compartment m to compartment n is a constant, independent of compartment stock levels.

(ii) $F_{mn} = \phi_{mn}x_m$: donor-controlled flow. ϕ_{mn} are analogous to birth and death rates in population models.

(iii) $F_{mn} = \phi_{mn}x_n$: recipient-controlled flow. ϕ_{mn} are analogous to the input–

output coefficients in economic models. Patten cites the case of herbivores when plants are in abundant supply.

(iv) $F_{mn} = \phi_{mn}x_m x_n$: donor–recipient controlled flow. This could be analogous to migration rates (if they are modelled separately with a spatial interaction model) in a population model, but caution has to be exercised here because of the quadratic nature of the $x_m x_n$ term. (In the migration model ϕ_{mn} would have the dimension $[x_m]^{-1}$.)

(v) $F_{mn} = \phi_{mn}x_m(1 - \alpha_{mn}x_n)$: donor-controlled modified by negative feedback from the recipient.

(vi) $F_{mn} = \phi_{mn}x_m(1 - \alpha_{mn}x_m)$: donor-controlled, modified by negative feedback from donor.

(vii) $F_{mn} = \phi_{mn}x_n(1 - \alpha_{mn}x_n)$: recipient-controlled, with added recipient-size negative feedback.

(viii) $F_{mn} = \phi_{mn}x_n(1 - \alpha_{mn}x_m - \beta_{mn}x_n)$: recipient-controlled, with negative feedbacks related to both donor and recipient populations. This is essentially the Lotka–Volterra mechanism which will be pursued further in another context in Chapter 8.

For completeness, it is tempting to add a ninth:

(ix) $F_{mn} = \phi_{mn}x_m(1 - \alpha_{mn}x_m - \beta_{mn}x_n)$: donor-controlled, modified by Lotka–Volterra negative feedbacks.

The first three mechanisms are linear; the remaining six nonlinear. When these substitutions are made for flows in accounting equations (5.28), simultaneous (possibly nonlinear) difference or differential equations are generated which describe the dynamics of the system. It is easy, then, to simulate change numerically. For illustration, take equation (5.27) and suppose the flows are replaced by type (ii) mechanisms from the above list, say as

$$x_{mn}^k = \phi_{mn}^k S_m^k \qquad (5.88)$$

and

$$x_{en}^k = \phi_{en}^k E^k \qquad (5.89)$$

(where E^k is the environmental 'stock'). Equation (5.27) then becomes

$$S_n^k(t + \delta t) = \sum_m \phi_{mn}^k S_m^k(t) + \phi_{en}^k E^k \qquad (5.90)$$

and the time sequence can be simulated straightforwardly.

5.4.7 Example 5: towards a water pollution model

In Section 5.2.4, we outlined the form of accounting equation used for animal or plant volumes by Kelly and Spofford. This forms the basis of a very ambitious model, the details of which would take us beyond the scope of this section. Here, we attempt to outline the principles involved.

First, we use the label i for 'reach' and make this explicit in equation (5.33) which now becomes

$$x_{imt+\delta t} - x_{imt} = f_{im} - r_{im} - d_{im} - e_{im} - p_{im} + y_{im} - wx_{imt}. \qquad (5.91)$$

The flows f_{im}, r_{im}, d_{im}, and e_{im} all represent biological processes which can be set as rates multiplied by x_{im}^t. (Excretion is taken by Kelly and Spofford as a constant multiplied by $(x_{im}^t)^2$ and this gives the first group of terms a logistic shape: $Ax_{imt} - Bx_{imt}^2$, for suitably defined A and B.) The predator rates are also biological and depend, nonlinearly, on the other populations. The flows y_{im} depend on the assumptions to be made about interaction with higher reaches, together with data about industrial effluent and any other exogenous inputs.

The rates, as defined in this way, depend on the volumes of dissolved oxygen and nutrients which are given by other reach-compartment equations, and these—and the animal/plant rates—depend on the final three properties: turbidity, toxicity, and temperature. All the relationships involve nonlinear curves fitted to empirical data, and this makes the overall structure of the model very complicated indeed.

Once the individual rate-functions are specified the model can be pushed sequentially through time using the method described for the compartment model in the previous subsection. All the variables on the right-hand side refer to time t and the increment for each main state variable can be calculated. The overall behaviour will be very complicated due to interdependence and nonlinearities, but Kelly and Spofford report quite good fits to empirical data.

The purpose of the model is to help in pollution control within a mathematical programming framework, and this is pursued further as an example in Chapter 10. Aspects of this kind of dynamic behaviour will also be considered further in another context in Chapter 8.

5.5 MORE COMPLICATED ACCOUNT-BASED MODELS

5.5.1 How more complicated models arise

The models outlined in the previous section were based on rate definitions which were directly associated with accounting elements. The need for more complicated models arises in two ways, which are sometimes related. First, there is often an acute shortage of data in relation to many accounting elements, which means that many of the rates cannot be measured. Thus new procedures have to be adopted to fill the information gaps. Secondly, the modeller may wish to use more aggregate rate definitions than would be implied by the accounting array, especially when it is a high dimensional one such as $\{x_{ij}^{mnk}\}$. Indeed we saw an example of that at the end of the previous section, though we kept that model in the simple category because the rates were directly related to aggregating accounting arrays $\{U^{kn}\}$ and $\{V^{mk}\}$.

It will already be clear that, with more complicated models, few general rules apply, and so we proceed largely by example. First, however, we comment on the role of balancing factors, which arise from entropy maximizing methods in filling in data gaps.

5.5.2 Balancing factors

Balancing factors, it will be recalled, are terms such as A_i and B_j which appear in the doubly constrained spatial interaction model introduced in Chapter 4. It turns out that those factors have a more general use in helping to fill gaps in accounting (and indeed any other) matrices. At least three kinds of problem can be identified: first, elements of the accounts are missing; secondly, the available data is not sufficiently disaggregated; and thirdly, an accounting matrix is known but is inconsistent with other known information such as row and column totals. Chilton and Poet (1973) provide useful examples to illustrate the first two types and then we turn to input–output modelling in relation to the third.

The *Census of Population* for 1966 contains accounts of migration flows, $\{K^{ij}\}$ say, between London Boroughs. This is a ten percent sample, and in order to avoid infringing confidentiality, an entry is left blank if it contains less than 10 people. How can the research worker estimate the missing entries? Let P_i and Q_j be the known totals of the ith and jth column respectively, and let O_i and D_j be the totals of the published entries. The shortfalls are

$$X_i = P_i - O_i \tag{5.92}$$

and

$$Y_j = Q_j - D_j \tag{5.93}$$

say. Let γ denote the set of (i, j) pairs for which the published entry is blank. Then, we will use the notation $(i, j) \in \gamma$ to denote the (i, j) pair being a member of this set ('$\in$' being the set inclusion sign). We then know that, for $(i, j) \in \gamma$,

$$\sum_{js.t.(i,j) \in \gamma} K_{ij} = X_i \tag{5.94}$$

and

$$\sum_{is.t.(i,j) \in \gamma} K_{ij} = Y_j \tag{5.95}$$

(where the summations have to be carefully defined). A 'balancing factor' estimate of the missing T_{ij}'s then turns out to be

$$K_{ij} = A_i B_j \qquad (i, j) \in \gamma \tag{5.96}$$

where

$$A_i = X_i / B_j \tag{5.97}$$

$$B_j = Y_j / A_i. \tag{5.98}$$

Equations (5.97) and (5.98) can be solved iteratively (in a similar way to the A_i's and B_j's in the doubly constrained spatial interaction model).

The second type of problem can be illustrated from similar data. Suppose we need $\{K_r^{ij}\}$, migrants from i to j, aged r at the beginning of the period. Published data offers not this, but $\{K^{ij}\}$, $\{O_r^i\}$, and $\{D_r^j\}$ where

$$\sum_j K_r^{ij} = O_r^i \tag{5.99}$$

$$\sum_i K_r^{ij} = D_r^j \tag{5.100}$$

and

$$\sum_r K_r^{ij} = K^{ij}. \tag{5.101}$$

Then a balancing factor estimate of $\{K_r^{ij}\}$ is

$$K_r^{ij} = A_r^i B_r^j C^{ij} \tag{5.102}$$

where

$$A_r^i = \frac{O_r^i}{\sum_j B_r^j C^{ij}} \tag{5.103}$$

$$B_r^j = \frac{D_r^j}{\sum_i A_r^i C^{ij}} \tag{5.104}$$

and

$$C^{ij} = \frac{K^{ij}}{\sum_r A_r^i B_r^j} \tag{5.105}$$

and again these equations can be solved iteratively.

The third kind of problem arises when there is a prior matrix but which does not conform to other known information such as row and column totals. Suppose, for example, that we wish to construct economic accounts Z_B^{mn} for region B but all we have available is a set of accounts Z_A^{mn} for another region A (but which is thought to be 'similar') and known row and column totals for region B, say $\{F^m\}$ and $\{G^n\}$. Then, Z_A^{mn} can be 'adjusted' by balancing factors $\{R^m\}$ and $\{S^n\}$ to ensure that

$$\sum_n Z_B^{mn} = F^m \tag{5.106}$$

and

$$\sum_m Z_B^{mn} = G^n. \tag{5.107}$$

This is accomplished by

$$Z_B^{mn} = R^m Z_A^{mn} S^n \tag{5.108}$$

where

$$R^m = \frac{F^m}{\sum_n Z_A^{mn} S^n} \tag{5.109}$$

and

$$S^n = \frac{G^n}{\sum_m R^m Z_A^{mn}} \tag{5.110}$$

and the last pair of equations can be solved iteratively in the usual way. This is the much-used 'RAS' method — so-called because of the original notation adopted in its input–output modelling use.

5.5.3 Balancing factors and entropy maximizing

The adjustment methods described above can easily be seen to do what was required, but they can be closely connected to entropy maximizing methods. This is useful because it establishes this method of adjustment (for there are alternatives) as being the 'least biased with respect to known information' in the usual entropy maximizing sense and it also provides a method for handling more complicated problems which sometimes arise (and we will have an example of such a problem in Section 5.5.5).

The general method relies on the Kullback entropy introduced in Chapter 4. Let $\hat{T}_{ij\ldots}$ be a prior estimate of some array (where the dots following the subscripts i and j indicate that any number of indices can be involved) and let $T_{ij\ldots}$ be the revised estimate. Then this is obtained by maximizing.

$$S = - \sum_{ij\ldots} T_{ij\ldots} \, \log\left(\frac{T_{ij\ldots}}{\hat{T}_{ij\ldots}}\right) \tag{5.111}$$

subject to any known constraints.

If no prior estimate, $\hat{T}_{ij\ldots}$, is available, then that term in (5.111) is set to 1 and the usual entropy term is maximized.

We can now see how the three examples of the previous section can be derived in this way. Equations (5.96)–(5.98) arise from

$$\text{Max } S = - \sum_{(i,j)\in\gamma} K^{ij} \log K^{ij} \tag{5.112}$$

such that

$$\sum_{j \text{ s.t.} (i,j)\in\gamma} K^{ij} = X^i \tag{5.113}$$

and

$$\sum_{i \text{ s.t.} (i,j)\in\gamma} K^{ij} = Y^j. \tag{5.114}$$

Equations (5.102)–(5.105) arise from

$$\text{Max } S = - \sum_{ijr} K_r^{ij} \log K_r^{ij} \tag{5.115}$$

such that

$$\sum_j K_r^{ij} = O_r^i \tag{5.116}$$

$$\sum_i K_r^{ij} = D_r^j \tag{5.117}$$

and

$$\sum_r K_r^{ij} = K^{ij}. \tag{5.118}$$

In each of these two examples, there is no prior matrix. In the third, there is one: equations (5.108)–(5.110) arise from

$$\text{Max } S = - \sum_{mn} Z_B^{mn} \log \left(\frac{Z_B^{mn}}{Z_A^{mn}} \right) \tag{5.119}$$

such that

$$\sum_{n} Z_B^{mn} = F^m \tag{5.120}$$

and

$$\sum_{m} Z_B^{mn} = G^n. \tag{5.121}$$

5.5.4 A population example

We now return to the full population accounting array defined by equation (5.20). It is represented in Figure 5.5 with row and column totals shown, and the shaded areas of the figure show the items which are usually available from data. For convenience, we neglect age subscripts for the time being. The problem facing the model builder is obvious: most of the elements of the array are not available from data. This means that the usual rates cannot be calculated directly, and this problem provides the nub of the solution. This is to calculate rates based on 'at-risk' populations and then no use these rates to estimate many of the missing elements in the accounts. The method is only sketched here, as it is presented in detail elsewhere.

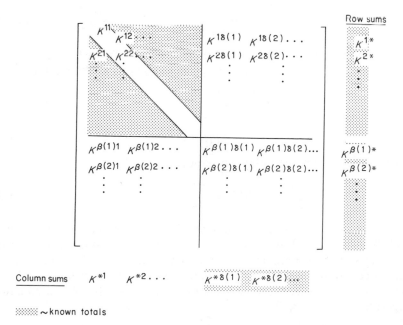

Figure 5.5 An array of spatial population variables

The population at risk of dying, for example, in region i during a period t to $t + T$ is made up of proportions of all the accounting elements which involves spending some time in i. The notation used is $\hat{K}^{D*i}$ for this quantity and it can be written

$$\hat{K}^{D*i} = \sum_{jk} {}^i\theta^{Djk} K^{jk} \tag{5.122}$$

where the summation is over all states (including β- and δ-states) and ${}^i\theta^{Djk}$ is the proportion of time someone in the (j, k) flow spends in region i during the period. For example, in the migrant term ${}^i\theta^{Dik} K^{ik}, {}^i\theta^{Dik}$ would normally be taken as 0.5 to indicate that *on average* (Weaver-II again) half the time is spent in i and half in k. Suppose now that this at-risk quantity can be calculated; this is the population which matches observed (and therefore available) total deaths in i, $K^{*\delta(i)}$ and so a death rate d^i can be calculated as

$$d^i = K^{*\delta(i)}/\hat{K}^{D*i}. \tag{5.123}$$

So in this case, by defining an at-risk population, we have been able to calculate a rate using a known row total rather than an accounting element directly.

This rate, however, is assumed to apply within accounting elements and can be used to estimate them. All the so-called minor flows are calculated in this way. For example

$$K^{i\delta(j)} = d^j \hat{K}^{Di\delta(j)} \tag{5.124}$$

since, within j, we assume

$$d^j = \frac{K^{i\delta(j)}}{\hat{K}^{Di\delta(j)}} \tag{5.125}$$

where $\hat{K}^{Di\delta(j)}$ is the at-risk population for the $(i, \delta(j))$ flows. To see how the method then works out very neatly, take as an approximation to $\hat{K}^{Di\delta(j)}$ the following:

$$\hat{K}^{Di\delta(j)} = K^{ij} + K^{i\delta(j)}. \tag{5.126}$$

So, from (5.124),

$$K^{i\delta(j)} = d^j(K^{ij} + K^{i\delta(j)}) \tag{5.127}$$

which can be solved for the unknown term $K^{i\delta(j)}$ as

$$K^{i\delta(j)} = \frac{d^j K^{ij}}{1 - d^j}. \tag{5.128}$$

Birth rates are handled similarly and, in all, the missing terms $\{K^{i\delta(j)}\}$, $\{K^{\beta(i)j}\}$, and $\{K^{\beta(i)\delta(j)}\}$ (in the last case including the (i, i) term, in the first two cases not) are handled in this way. The remaining missing 'major' terms are calculated from the accounting equations, making use of known row and column information as follows. First, estimate $K^{i\delta(i)}$, for each i, from the $(N + i)$th column equation (see Figure 5.5):

$$K^{i\delta(i)} + \sum_{j \neq i} K^{j\delta(i)} + K^{\beta(i)\delta(i)} + \sum_{j \neq i} K^{\beta(j)\delta(i)} = K^{*\delta(i)} \tag{5.129}$$

Then estimate K^{ii} from the ith row equation:

$$K^{ii} + \sum_{j \neq i} K^{ij} + K^{i\delta(i)} + \sum_{j \neq i} K^{i\delta(j)} = K^{i*}. \tag{5.130}$$

Then estimate $K^{\beta(i)i}$ from the $(N + i)$th row equation:

$$K^{\beta(i)i} + \sum_{j \neq i} K^{\beta(i)j} + K^{\beta(i)\delta(i)} + \sum_{j \neq i} K^{\beta(i)\delta(j)} = K^{\beta(i)*}. \tag{5.131}$$

This completes the accounts, and the ith column equation can then be used to estimate the now final populations K^{*i}:

$$K^{ii} + \sum_{j \neq i} K^{ji} + K^{\beta(i)i} + \sum_{j \neq i} K^{\beta(j)i} = K^{*i}. \tag{5.132}$$

The only problem with this procedure is that the at-risk population, such as that in (5.122), *depends on the missing flows*. This turns out to be only a computational problem, however: the equations are solved iteratively. At-risk populations and rates are calculated first assuming that all the minor flows are zero (with major flows estimated from the accounting equations); then minor flows are calculated; the rates revised; and so on.

5.5.5 A new input–output model

We now explore a new way, due to Macgill (1977a) of getting more information into an input–output model. We continue the argument from the end of Section 5.4.4 and use the same variables. The main array is taken as $\{x^{mnk}\}$ though it is recognized that the data will be an absorption matrix $\{U^{kn}\}$ and a make matrix $\{V^{mk}\}$. A suitable estimate of x^{mnk} is then

$$x^{mnk} = \frac{V^{mk} U^{kn}}{Z^k} \tag{5.133}$$

where

$$Z^k = \sum_m V^{mk} = \sum_n U^{kn}. \tag{5.134}$$

This in fact arises from maximizing an entropy function

$$S = - \sum_{mnk} x^{mnk} \log x^{mnk} \tag{5.135}$$

subject to

$$\sum_n x^{mnk} = V^{mk} \tag{5.136}$$

and

$$\sum_m x^{mnk} = U^{kn} \tag{5.137}$$

and so is another example of entropy maximizing being used to fill in gaps in data. This is a base year estimate of $\{x^{mnk}\}$ which we label as $\{\bar{x}^{mnk}\}$. A difficult question is how to forecast with input–output models. The simple model would involve an assumption about future final demand, $\{Y^m\}$, and the constancy of the technical coefficients. By using entropy maximizing again, it is possible to be slightly less restrictive and to use one set of coefficients, the absorption coefficients b^{kn} (formed as U^{kn}/X^n) and to rely on $\bar{x}^{mnk}$ to provide prior information on the structure of the economy in a Kullback entropy function. The model is then

$$\text{Max } S = - \sum_{mnk} x^{mnk} \log\left(\frac{x^{mnk}}{\bar{x}^{mnk}}\right) \tag{5.138}$$

such that

$$\sum_m x^{mnk} = b^{kn}\left(\sum_{m'k'} x^{nm'k'} + Y^n\right). \tag{5.139}$$

This provides an alternative to the rectangular model.

5.5.6 Other possibilities

It is clear that the principle involved in building new models is to use accounting relations, when they are known, as constraint equations (and to add any other known information) and then to maximize an entropy function (possibly of the Kullback type including prior probabilities) to estimate the state variables. Here we simply remark that, although it does not seem to have been attempted, there is no reason why this should not be done with ecological or ecological-water-resource models such as those outlined in Section 5.4.

5.6 CONCLUDING COMMENTS

Accounting, as we have seen, provides the methods for keeping track of all system components and so ensuring both *comprehensiveness* and *consistency*. The former attribute connects to our Chapter 2 discussion of definition of state variables, through the definition of component states, the counting of components in these states, and hence a definition of the system state. In appropriate cases, this provides, almost directly, an adequate basis for system modelling using rates of change.

The second attribute, consistency, often manifests itself in the form of *constraint* equations. These can be used directly, as in the population model of Section 5.5.4, as model equations; or in the cases where much information in the accounting array is missing, but some 'total' information is available, they provide constraint equations for entropy maximizing equations. Many of these constraints, certainly the 'keeping track' ones, are essentially the *conservation* equations of the models and as such play a role in a wide variety of models. But other types of constraints equations arise from observations of accounting arrays. For example, with an $\{x^{mnk}\}$ array, if k is some conserved commodity, so that inflows and outflows of that commodity at a zone must balance, we get *materials balance constraints*.

Finally, we note that because accounts are concerned with change, they often form the basis of dynamic models. This point will be taken up again in Chapter 8, but a couple of preliminary points will set part of the scene. Consider the array $\{K_{pq}\}$ of Section 5.1 where p and q are typical state labels. Then

$$K_{p*}(t) = \sum_q K_{pq}(t, t + T) \tag{5.140}$$

and

$$K_{*q}(t + T) = \sum_p K_{pq}(t, t + T). \tag{5.141}$$

Subtracting the first of these equations from the second (and putting $q = p$ and $p = q$ in the second) gives

$$K_{*p}(t + T) - K_{p*}(t) = \sum_{q \neq p} K_{qp}(t, t + T) - K_{pq}(t, t + T). \tag{5.142}$$

This is the change in number of occupants of state p and provides the basis of *difference equations*. By dividing by T and letting $T \to 0$, it provides the basis of differential equations also. (We gave a particular illustration of this principle with the ecological model of Section 5.4.5.)

One problem with this account-based approach to dynamics arises if p is a rather long subscript list. For one example which has arisen, for example, p is (i, j, k, w) which is (residence location, job location, house type, and job type). The number of people in a state at time t is $T_{ij}^{kw}(t)$, say, and a corresponding accounting variable is $K_{iji'j'}^{knk'n'}(t, t + T)$, which is an eight-dimensional array. In such cases, rather than attempting to model the account elements directly, it is assumed that movers go into a 'pool' and then out again, as shown in Figure 5.6. The modelling task is then to estimate two four-dimensional arrays giving the transfers from (i, j, k, w) to the pool, and from the pool to (i', j', k', w'). This is obviously simpler, though of course information is lost also — individuals in particular origin states cannot be associated with corresponding individuals in final states.

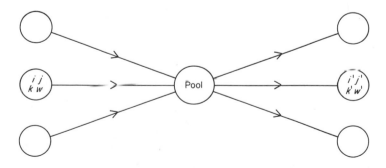

Figure 5.6 A mover pool

NOTES ON FURTHER READING

Many of the examples in this chapter are based on population or economic accounts. Useful general background references in this respect are Rees and Wilson (1977, †Chapters 1–3, *other chapters) and *Stone (1966, 1970) respectively. The population models in this form originate in the work of †Leslie (1945) and were given a spatial dimension by †Rogers (1966). The economic models originate in the work of Leontief in the 1940's (see, for example, †Leontief, 1951, 1967).

It is also useful to cite some references which form more specific background to some particular subsections. A variety of models of social mobility are presented by *Bartholemew (1970) (5.4.2). The classic example of the urban input–output models is that of *Artle (1959) (5.4.4). The input–output model is linked to ecological sectors by *Isard et al. (1972). *Patten (1971) provides a useful primer on ecological modelling (5.4.6) and the subsection of inter-reach accounting is based on the work of †Kelly and Spofford (1977). A useful review of balancing factors and some related theorems is to be found in the work of *Macgill (1975, 1977b) (5.5.2 and 5.5.3) who also offers extended approaches to input–output modelling (*Macgill, 1977a) (5.5.5). A specific application of the entropy concept to finding missing data in accounts appears in *Chilton and Poet (1973). In Section 5.6, we mention materials balance constraints—another form of accounting equations—and these can be pursued in *Cripps, Macgill, and Wilson (1974).

CHAPTER 6

Optimization

6.1 THE NATURE OF OPTIMIZATION

In some systems, one or more of the state variables may be determined by an *optimizing process*. Three very simple examples are shown in Figure 6.1. The first, (a), shows a graphical representation of how an individual might decide how many hours to work. It is assumed that he maximizes his utility, which will itself be a function of income derived from work, leisure time, sleeping time, and so on. It is easy to see that the curve is likely to take the form shown in Figure 6.1 (a) and that H_{opt} is the value of H for which U is a maximum. It is not argued that individuals literally operate in this way, of course, but that such a graph provides a model of their behaviour.

The next example, (b), shows the process whereby the manager of a firm might decide how much to invest. He seeks to maximize profit per employee. And in the third example, (c), a water authority, which knows, let us assume, that it wants to build a reservoir on a certain site, determines the size of that reservoir by finding S_{opt} to minimize the cost, C, per unit of water supplied. Note that optimization can involve maximization or minimization in different circumstances. Note that in these examples, the manager and the water authority may *literally* operate this way by asking their technical staff, say engineers and economists, to construct such curves for them.

In each of these cases, the function being maximized amounts to a representation of the *process* involved, and this determines the state variable. Biological or physical processes can also be described in this way. Figure 6.1(d) shows the water intake of a species in an ecosystem to be determined by the volume needed to maximize its biomass. Finally, Figure 6.1(e) shows the channel width carved by a river with a particular flow being determined by the minimization of energy loss in the water transport process.

These are all simple examples of optimization processes, at least in the sense that very few state variables are involved. If the process function can be specified in terms of the state variables, then finding the optimum value can be achieved by plotting the curve, as indicated in the examples of Figure 6.1, or more precisely and formally by using standard methods of elementary calculus. Further, these

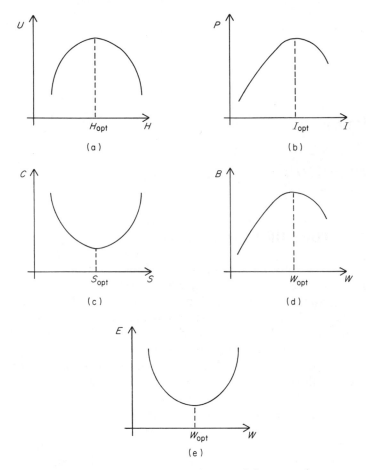

Figure 6.1 Examples of maxima or minima as optima

examples are all *unconstrained* problems; we will see shortly that much complexity is introduced via constraints.

The main point of connection between optimization and the Chapter 2 discussion of systems concepts can now be seen as related to the *processes* which determine some or all of the state variables. When such processes can be represented by optimization mechanisms, the methods of modelling to be described in this chapter will be appropriate. Note that 'optimization' here is being used in a technical sense — describing the outcome of a process in which something is maximized or minimized — and does not necessarily imply 'for the best' in some value sense (though it does not exclude that either). More complexity arises in practice from subsystem structures. An optimization process will often refer to a subsystem and the whole system, indeed, may then consist of several interacting optimization processes. An example of this is where a government authority is attempting to locate a facility in some optimum way in the knowledge that, within the whole system, users will determine the way in which they use the facility according to their own optimization (say utility maximizing) processes. In such a case, there is 'optimization within optimization'.

In some cases, we will see that optimization mathematics is being used to solve aggregation problems for Weaver-II systems; in others, the complexity arising from subsystem interaction means that the overall system behaviour is Weaver-III and that the methods offered here offer one way of tackling such problems.

6.2 AN INFORMAL INTRODUCTION TO OPTIMIZATION MATHEMATICS

6.2.1 Objective functions and constraints

The function which is to be optimized is called the *objective function*. Typically, it is to be optimized subject to constraints. Suppose, for example, a manager of a firm has to invest in facilities for two kinds of products — amounts of investment x and y respectively. The profit he expects is

$$P = 100x + 300y \qquad (6.1)$$

but he has a total budget B, so that

$$x + y \leqslant B \qquad (6.2)$$

is the budget constraint. Also, each unit of investment x employs two workers, and each unit of y one worker; that is the first product is more labour intensive though less profitable. He has an agreement with a union that the minimum number of jobs to be created is E, so this requires

$$2x + y \geqslant E. \qquad (6.3)$$

The first product sells in his shops at half the price, per unit of investment, of the other: 20 as against 40. To keep the shops viable, he has to create a minimum additional revenue of S. Thus

$$20x + 40y \geqslant S. \qquad (6.4)$$

x and y can now be found from an optimization problem which consists of maximizing P in equation (6.1) subject to constraints (6.2)–(6.4). This can be written briefly as

$$\text{Max}_{(x,y)} P = 100x + 300y \qquad (6.5)$$

subject to

$$x + y \leqslant B \qquad (6.6)$$

$$2x + y \geqslant E \qquad (6.7)$$

$$20x + y \geqslant S \qquad (6.8)$$

together with non-negativity restrictions on x and y $(x, y \geqslant 0)$. This kind of optimization problem, involving an objective function in several variables, and constraints, is known as a mathematical programming problem. This particular example is a *linear programming problem* because the objective function and the constraints are all linear in the variables x and y.

A two-variable problem of this type can be interpreted graphically, and this gives some insight which can be carried over to more complicated problems involving more variables. First, consider equations (6.6)–(6.8) in turn with equality signs in place of inequality signs, and definite numerical values for B, E, and S. The equations represent three lines which can be plotted as shown in Figure 6.2. The effect of the constraint inequality signs can then be represented by shading the *feasible* side of the line. For example, with constraint (6.6), $x + y \leqslant B$ and points (x, y) satisfy this if they are below the line; similarly for the other two constraints. It is then easy to see that the area for which all constraints are satisfied simultaneously is the triangle ABC which forms an enclosed 'convex' region.

Now consider the objective function (6.6). For a given value of P, this represents a straight line also and $P/100$ is the intercept on the y-axis. Thus the solution of the problem is to find a point on this line (which moves up and down as P varies) which is also inside the feasible region. A number of such lines are plotted on Figure 6.2, as dashed lines, for varying values of P. It can easily be seen that the point A is the optimal point: the line in position (1) does not pass through the feasible region at all; in positions (3), (4), ..., there are always 'better' positions giving higher P values, and the optimum is position (2) which passes through A.

This demonstrates one universal feature of linear programming problems — that the optimum solution always lies at a corner of the feasible region. (There is only one reservation to this statement. In certain circumstances, such a point is not unique. For example, if AB was parallel to the objective function line, A and B and any point on AB lying between them can be seen to be optimum points.)

This diagrammatic presentation was possible because only two variables were involved. Usually, programming problems involve large numbers of variables

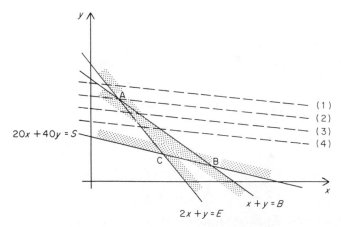

Figure 6.2 An example of linear programming

and constraints. A 100-variable problem, say with 50 constraints, would not be unusual. A geometrical presentation of this would involve a 100-dimensional space, the constraints would be hyper-planes, and they would intersect to form a polyhedral feasible region. The optimum solution would still be a point at a corner of the feasible region. We will present a number of examples illustrating different types of linear programming problem in the next section.

6.2.2 Types of linear programming problem

In the example presented above, the variables were allowed to vary continuously. Sometimes the problem only makes sense if variables can take integer values. A problem which is wholly made up of variables of this kind is called an *integer programming problem*. When a variable can vary continuously, it is said to take a range of real values, and a problem made up wholly of such variables can be called a *real* programming problem; but 'real' is usually 'assumed' in such a case and only used in the mixed case, where some variables are integer, some continuous, and these are called *mixed integer–real programming problems*.

The example in the previous section was a real problem. An example of a pure integer problem would arise if we wished to locate a number of facilities as j, say n_j, in such a way that costs were minimized but subject to a number of constraints. It is assumed that the cost of each facility varies with j and is designated c_j. This problem could be formulated as

$$\text{Min } Z = \sum_{j=1}^{N} c_j n_j \tag{6.9}$$

subject to

$$\sum_{j=1}^{N} a_{ij} n_j \leqslant b_i, i = 1, \ldots, M, n_j \text{ integer.} \tag{6.10}$$

We have assumed N zones in all, and note the economical way in which the M constraints have been written. A 'less than' sign has been used without loss of generality because some of the coefficients a_{ij} or the constants b_i could be negative. Finally, we have added the condition that each n_j must be an integer.

An example of such a problem would be the siting of factories, n_j in j at cost c_j each, in such a way as to maintain pollution levels in residential areas i below some minimum, $b_i \cdot a_{ij}$ would be the average pollution impact of a factory in j on area i, and would be assumed given. The variables in such problems are sometimes confined to the values 0 or 1 if there is a maximum of one facility per zone.

An ingenious formulation of a facility location problem is due to Revelle (1968), cited by Scott (1971). Again, we have a spatial system and zones labelled, $1, 2, \ldots, i, j, \ldots, N$. M facilities are to be located. Let r_i be the demand for use of such a facility in residence zone i and let d_{ij} be the distance from i to j. Then the main variable is λ_{ij} which is 1 if the residents of i are allocated to the facility in j and 0 otherwise. The ingenuity is displayed in the way the constraints are defined.

The problem is based on distance minimization for users:

$$\text{Min } Z = \sum_{i=1}^{N} \sum_{j=1}^{N} r_i \lambda_{ij} d_{ij} \quad \text{(6.11)}$$
$$\{\lambda_{ij}\}$$

subject to

$$\sum_{j=1}^{N} \lambda_{ij} = 1 \quad \text{(6.12)}$$

(which ensures that the residents of i are assigned to one and only one facility).

$$\sum_{j=1}^{N} \lambda_{jj} = M \quad \text{(6.13)}$$

(which ensures that M facilities are located in all).

$$\lambda_{jj} - \lambda_{ij} \geqslant 0, \quad i = 1, 2, \ldots, N, \quad \text{(6.14)}$$
$$j = 1, 2, \ldots, N,$$
$$i \neq j$$

(which ensures that if λ_{jj} is zero, and there is no facility at j in the optimum solution, then λ_{ij} cannot be 1 and no residents of i are assigned to it). Finally, there are the non-negativity constraints

$$\lambda_{ij} \geqslant 0. \quad \text{(6.15)}$$

This is an example of a *location–allocation* problem, since the location of facilities *and* their pattern of use is determined.

A mixed integer–real problem would arise in facility location if there were two types of facilities to be located, one involving fixed costs at a location and the other involving costs dependent on size. Then, there would be two sets of state variables: r_j, the number of facilities of the first kind, and x_j, the size of the second kind of facility in j. An objective function (given costs p_j and c_j respectively) would be

$$\text{Min } \sum_i (p_i n_j + c_j x_j) \quad \text{(6.16)}$$
$$\{n_j, x_j\}$$

subject to an appropriate set of constraints. (If there was any interdependence between the two types of facilities, for example if the second type j could only exist if the first was already located in a zone, then this could be represented in the constraints but the problem would become nonlinear.)

As a third example, we take a pure-real problem—the well known and useful *transportation problem of linear programming*. This makes use of the same variables as the basic entropy-maximizing journey to work example, and the two problems have an interesting relationship as we shall see in Section 6.6.1. Consider $\{T_{ij}\}$ to be some general array of interaction variables, and $\{O_i\}$ and $\{D_j\}$ to be given row and column totals; let $\{c_{ij}\}$ be the interaction cost matrix. Then an interesting problem is to find $\{T_{ij}\}$ such that

$$\text{Min } Z = \sum_{ij} T_{ij} c_{ij} \qquad (6.17)$$
$$_{\{T_{ij}\}}$$

subject to

$$\sum_j T_{ij} = O_i \qquad (6.18)$$

and

$$\sum_i T_{ij} = D_j. \qquad (6.19)$$

This set of T_{ij}'s answers the question: how do we ship goods (or whatever) from i to j, given supplies O_i at each i and demands D_j at each j? An example of this problem would be if the origins were petroleum refining points (say located near oil fields) and the destinations were petrol consuming points. The solution $\{T_{ij}\}$ would give the least-cost way of shipping between the two kinds of points.

6.2.3 Nonlinear programming

The functions used to illustrate unconstrained optimization problems in Section 6.1 (see Figure 6.1) were clearly nonlinear. Constraints too could be nonlinear. If *any* function appearing in a mathematical programming problem, either in the objective function or the constraints, is nonlinear, then the whole problem is said to be nonlinear. Such problems have to be sharply distinguished from linear ones because the powerful solution procedures developed for the latter are no longer applicable. This is easy to see from the geometrical example of the previous subsection: if the objective function had been nonlinear, then it would not *necessarily* be true that the optimum would be at a corner of the feasible region. For nonlinear problems, a number of cases can be distinguished, and general solution procedures are available for some of them. These points will be picked up in the subsection on solution procedures below and in a number of the examples which follow.

6.2.4 Duality

Any mathematical programme has associated with it a *dual* programme. In the dual, there is a variable in place of each constraint in the primal (which is what the original problem is called) and vice versa. Except in very special circumstances, the dual of the dual is the original primal, as one would expect.

The concept of duality can be first illustrated by the transportation problem of linear programming. The primal problem was given in equations (6.14)–(6.19) above. The corresponding dual problem is

$$\text{Max } Z' = \sum_i \alpha_i O_i + \sum_j v_j D_j \qquad (6.20)$$
$$_{\{\alpha_i, v_j\}}$$

such that

$$c_{ij} - \alpha_i - v_j \geqslant 0. \tag{6.21}$$

Note that the dual problem is a maximization problem when the primal is a minimization one (and vice versa). The dual variables are the vectors $\{\alpha_i\}$ and $\{v_j\}$ and they arise from the constraints (6.18) and (6.19) in the primal. There is a dual constraint, (6.21), arising from each primal variable, T_{ij}. With these constraints, the equality sign holds when $T_{ij} \neq 0$ and strict inequality when $T_{ij} = 0$.

It can be shown that, at the optimum, the primal and dual objective functions are equal:

$$Z = Z'. \tag{6.22}$$

The dual variables α_i and v_i can be interpreted as the 'comparative advantage' of locations i and j with respect to the associated constraints. Thus, (6.22) shows that minimizing total transport cost is equivalent to maximizing this concept of comparative advantage: dual variables hence often play an important role in locational economics and geographical theory.

In the general case, if the primal problem is written

$$\text{Max } Z = \sum_i a_i x_i \tag{6.23}$$

subject to

$$\sum_j b_{ij} x_j \leqslant c_j \tag{6.24}$$

then the dual is

$$\text{Min } Z' = \sum_j c_j y_j \tag{6.25}$$

subject to

$$\sum b_{ji} y_i \geqslant a_i. \tag{6.26}$$

It is slightly more difficult to construct nonlinear duals. One formulation is due to Balinski and Baumol (1968) and can be stated as follows. If the primal problem is

$$\text{Max } Z = f(x_1, x_2, \ldots, x_n) \atop \{x_i\} \tag{6.27}$$

such that

$$g_i(x_1, x_2, \ldots, x_n) \leqslant c_i \tag{6.28}$$

$$x_i \geqslant 0 \tag{6.29}$$

then the dual is

$$\text{Min } Z' = f + \sum_i v_i [c_i - g_i(\mathbf{x})] - \sum_j x_j \left[\frac{\partial f}{\partial x_i} - \left(\sum_i v_i \right) \left(\frac{\partial g_i}{\partial x_j} \right) \right] \atop \{v_i\} \tag{6.30}$$

such that

$$\left(\sum_i v_i \right) \left(\frac{\partial g_i}{\partial x_j} \right) \geqslant \frac{\partial f}{\partial x_j} \tag{6.31}$$

$$v_i \geqslant 0. \tag{6.32}$$

This may be useful for reference, but we will use a more transparent formulation, based on Lagrangian methods, in particular examples to be discussed below.

6.2.5 Solution procedures

Most solution procedures for mathematical programming problems cannot be stated analytically. They are *algorithms* represented in computers. The exception is the method for certain nonlinear problems, which does give considerable theoretical insight, which we outline below. For the most part, however, what the reader needs to know is that for particular types of problem, solution procedures exist and that he can seek the appropriate computer programmes at his local computer centre.

There are a number of general solution procedures for *linear* programming problems of which the best known is perhaps the simplex method. There are special versions of these procedures for particular problems such as the transportation problem. The essence of the task is speed of solution and the corresponding size of problem which can be tackled in a reasonable time on a given computer. With current machines, very large problems can be tackled involving thousands of variables and constraints. There are also a number of key theorems which arise out of these procedures of which the most important is that *the number of nonzero variables at the optimum solution of a linear programming problem cannot exceed the number of independent constraints*. In the case of the transportation problem, for example, there are $2N-1$ independent constraints (one degree of freedom being lost because $\sum_i O_i = \sum_j D_j$) so that not more than $2N-1$ of the T_{ij} variables—and there are, of course, N^2 such variables in all—can be nonzero. For N as 100, only 199 out of 10 000 variables can be nonzero. This result has important consequences for model building, as we will see in Section 6.2.7.

We now turn to nonlinear programming and the case where an analytical procedure exists and is helpful. Consider a simpler version of the nonlinear programming problem given by (6.27)–(6.29)—with equalities instead of inequalities in the constraints and no non-negativity condition:

$$\underset{\{x_i\}}{\text{Max}} \, Z = f(x_1, x_2, \ldots, x_N) \tag{6.33}$$

such that

$$g_i(x_1, x_2, \ldots, x_N) - c_i = 0, \quad i = 1, \ldots, M \tag{6.34}$$

(and note that each constraint is written in the form 'something equals zero' and that we have specified the number of constraints explicitly as M). A Lagrangian multiplier, λ_i, $i = 1, 2, \ldots, M$, is associated with each constraint in turn, and a Lagrangian function is formed as:

$$L = f(x_1, x_2, \ldots, x_N) + \sum_i \lambda_i [g_i(x_1, x_2, \ldots, x_N) - c_i]. \tag{6.35}$$

It can then be shown that the solution of (6.33)–(6.34) is the same as the solution of the *unconstrained* problem

$$\text{Max}_{\{x_i, \lambda_i\}} L(x_1, x_2, \dots, x_N, \lambda_1, \lambda_2, \dots, \lambda_M). \tag{6.36}$$

Instead of the M constraints, we have M additional variables $\lambda_1, \lambda_2, \dots, \lambda_M$ and L is to be maximized as a function of all the x_i and these, as shown explicitly in the notation in (6.36).

This unconstrained problem can then be solved by the standard methods of multivariate calculus, and provided the functions f and $\{g_i\}$ are reasonable, analytical solutions can be obtained. The method will not be spelled out in detail as it takes us beyond the mathematical scope of this book, but it will be sketched briefly for the benefit of those readers who have some calculus available. The solution is obtained by setting the partial derivatives of L with respect to each x_i and λ_i to zero:

$$\frac{\partial L}{\partial x_i} = \frac{\partial f}{\partial x_i} + \lambda_i \frac{\partial g_i}{\partial x_i} = 0, \quad i = 1, 2, \dots, N \tag{6.37}$$

$$\frac{\partial L}{\partial \lambda_i} = g_i(x_1, x_2, \dots, x_N) - c_i = 0, \quad i = 1, 2, \dots, M. \tag{6.38}$$

The second set of equations, (6.38), are simply the original constraint equations, and this is the essence of the trick: the way the multipliers are introduced ensures that, at the optimum, the constraints are satisfied and the sum which forms the second term of (6.35) vanishes.

We can apply this method both to the transportation problem of linear programming and to the entropy maximizing model of Chapter 4 (which is a nonlinear model because entropy is nonlinear). We do this in the spirit of much of the rest of this book of 'talking with equations' in order to gain insights rather than as a detailed presentation of a rigorous mathematical argument which the reader has to follow in detail.

Consider the programme given by equations (6.17)–(6.19). Let α_i be the Lagrangian multipliers associated with (6.18) and v_j those with (6.19). Then the Lagrangian is

$$L = \sum_{ij} T_{ij} c_{ij} + \sum_i \alpha_i \left(O_i - \sum_j T_{ij} \right) + \sum_j v_j \left(D_j - \sum_i T_{ij} \right). \tag{6.39}$$

At the optimum,

$$L = \sum_{ij} T_{ij} c_{ij} = Z. \tag{6.40}$$

For the dual problem given by equations (6.20) and (6.21), if T_{ij} is taken as the Lagrangian multiplier associated with (6.27), the Lagrangian is

$$L' = \sum_i \alpha_i O_i + \sum_j v_j D_j + \sum_{ij} T_{ij} (c_{ij} - \alpha_i - v_j). \tag{6.41}$$

In this case

$$L' = \sum_i \alpha_i O_i + \sum_j v_j D_j = Z' \tag{6.42}$$

at the optimum. But it can easily be seen by inspection of equations (6.39) and (6.41) that, by rearranging terms,

$$L = L' \tag{6.43}$$

and hence

$$Z = Z' \tag{6.44}$$

so that we have proved the result stated in equation (6.22) above. We can now see that the choice of notation for Lagrangian multipliers in (6.39) above was not accidental: such multipliers are in fact the dual variables, showing that there is an intimate connection between Lagrangians and duality.

Now consider the nonlinear programme given by equations (4.9)–(4.12) of Chapter 4 which are repeated here for convenience

$$\underset{\{T_{ij}\}}{\text{Max}}\, S = \log\left(\frac{T!}{\prod_{ij} T_{ij}!}\right) \tag{6.45}$$

(where $T = \sum_{ij} T_{ij}$) such that

$$\sum_j T_{ij} = O_i \tag{6.46}$$

$$\sum_i T_{ij} = D_j \tag{6.47}$$

$$\sum_{ij} T_{ij} c_{ij} = C. \tag{6.48}$$

The Lagrangian for this problem can be written

$$L = S + \sum_i \alpha_i (O_i - \sum_j T_{ij}) + \sum_j v_j (D_j - \sum_i T_{ij}) + \beta (C - \sum_{ij} T_{ij} c_{ij}) \tag{6.49}$$

which shows that

$$L = S \tag{6.50}$$

at the optimum. This can be rearranged to give

$$L = -\sum_{ij} T_{ij} (\log T_{ij} + \alpha_i + v_j + \beta c_{ij}) + \sum_i \alpha_i O_i + \sum_j v_j D_j + \beta C + T. \tag{6.51}$$

The dual constraint equations are

$$\log T_{ij} + \alpha_i + v_j + \beta c_{ij} = 0 \tag{6.52}$$

and the objective function is (to be minimized)

$$S' = \sum_i \alpha_i O_i + \sum_j v_j D_j + \beta C + T. \tag{6.53}$$

Thus, L in the (6.51) arrangement shows that

$$L = S' \tag{6.54}$$

also at the optimum. It can also be checked that this construction of the dual from the Lagrangian is identical to that which could have been derived from the Balinski and Baumol definition given in Section 6.2.4.

We see, therefore, that Lagrangian methods provide the solution procedures for entropy maximizing methods, and we will use the insights gained in this section further in Section 6.2.7.

In order to find problems to which Lagrangian methods were applicable, we had to replace inequality constraints by the more restrictive equalities and to drop the non-negativity constraints. When such complications are re-introduced, a number of solution procedures are available for particular classes of problem. Lagrangian theory can be generalized as Kuhn–Tucker theory which is possibly more important for the theoretical insight it offers rather than for any related computational procedures, but which again would take us beyond the scope of this book. For certain problems, computer programmes do exist and are very valuable—for example, a problem with a nonlinear but convex objective function and wholly linear constraints can be solved, and this is particularly useful for one of the examples which follows.

Some nonlinear problems are further complicated by the presence of integer variables. In most such cases (and indeed in many nonlinear problems without such complications), no general solution procedures exist which guarantee that the optimum can be found. Quite often, however, good heuristic procedures exist which produce at least approximate solutions. One set of such procedures are known as *branch and bound* methods, for example.

6.2.6 Optimization through random utility theory

In order to show that there are approaches to optimization which involve neither standard methods of calculus not those of mathematical programming, we explore some of the basic ideas of random utility theory. These ideas are presented through a particular example: an individual (of an assumed type n and living in a location i—though we will not make these labels explicit) chooses optimum destination, j, for some purpose (say work) and mode, k, to travel there. So by setting up the problem in this way, we have switched level of resolution and now take a more micro viewpoint. This in itself turns out to be important when we compare the results with our other model building techniques as well as illustrating a new method.

His choice is assumed to be governed by attributes associated with j purely, such as wage level, with k purely, such as comfort, and with (j, k) in combination, such as travel time. These attributes can be listed in a list, or vector, as $(z_{1j}^{(1)}, z_{2j}^{(1)}, \ldots, z_{1k}^{(2)}, z_{2k}^{(2)}, \ldots, z_{1jk}^{(3)}, z_{2jk}^{(3)}, \ldots)$ which can be written $(\mathbf{z}_j^{(1)}, \mathbf{z}_k^{(2)}, \mathbf{z}_{jk}^{(3)})$ for short. Even more briefly, we can designate it $\mathbf{z}^{jk}$—a vector made up of the three subvectors. Assume that a function of these variables exists which is a measure of the *utility* the individual derives from the choice (j, k) and call it $u(\mathbf{z}^{jk})$. u is also sometimes written u_{jk} for short. If the z's were all clearly measured and perceived and the individual had a clear view of his utility function, then the choice made could be written.

$$(j, k)_{\text{opt}} = \{j, k \text{ s.t. } u(\mathbf{z}^{jk}) > u(\mathbf{z}^{j'k'}), \forall j', k'\}. \tag{6.55}$$

(where '$\forall$' stands for 'for all'). That is, the (j, k) is chosen for which $u(\mathbf{z}^{jk})$ is a maximum out of all possible (j, k) pairs.

This (j, k) pair could be identified by enumeration of all $u(z^{jk})$ values in suitably simple cases. However, this is rarely possible in practice for all sorts of reasons. Individual utility functions (i.e. personal preferences) vary and are not easy to measure, an individual's information is not likely to be perfect in relation to all possible (j, k)'s, and so on. The solution to these difficulties is to seek a *probabilistic* solution to the problem. There are various ways of achieving this. One is to assume that the utility function is known, but that behaviour is probabilistic— that there is only a *probability* that the individual chooses the optimum. Luce (1959) showed that, in such a case, the probability of the individual choosing (j, k) could be represented as

$$P_{jk} = \frac{v_{jk}(\mathbf{z}^{jk})}{\sum_{jk} v_{jk}(\mathbf{z}^{jk})} \tag{6.56}$$

where

$$v_{jk} = e^{\lambda u_{jk}} \tag{6.57}$$

and λ is a parameter. This formula often works well *provided the alternatives are carefully specified.*

The second method is to assume that the individual maximizes his utility but that the utility function depends on parameters $\alpha_1, \alpha_2, \ldots = \boldsymbol{\alpha}$ (say, in vector notation) which vary across the population. Except in special cases, this method proves mathematically intractable however.

The third method is the basis of what is called *random utility theory*. Assume that u_{jk} is a function of a mean, u_{jk}^*, and an additive random variable, ε_{jk}, so that

$$u_{jk} = u_{jk}^*(\mathbf{z}^{jk}, \boldsymbol{\alpha}) + \varepsilon_{jk} \tag{6.58}$$

(where, again, the possible parameters, $\boldsymbol{\alpha}$, are shown explicitly). Then the individual is assumed to maximize utility, but ε_{jk} varies across the population. This method turns out to be mathematically tractable in many cases (though we shall not present any details here) and also offers new insights into decision-making processes. Further, since it operates at a more micro level of resolution, it is possible to cast some light onto more aggregate entropy maximizing and programming models.

The basis of the mathematics involved will just be mentioned at this point so that the reader can begin to explore (elsewhere—see further reading!) and thereafter we will simply highlight some of the main results. But this basis does emphasize the difference between this optimization procedure and others discussed earlier. The argument turns on the individual's perception of u, given u_{jk}^* and the distribution of ε_{jk}. Let $\Gamma_{jk}(u, u_{jk}^*) du$ be the probability that the perceived surplus is in the range u to $u + du$. Here du can be thought of as an increment. Let $\theta_{jk}(u)$ be the probability that all other alternatives have utility less than u_{jk}. It is then necessary to 'add up' over a lot of possibilities using integration, though most of the argument can be followed without any knowledge of integral calculus. So, if

$$\theta_{jk}(u) = \prod_{j'k' \neq jk} \int_{-\infty}^{u} du' \Gamma_{j'k'}(u', u_{j'k'}^*) \tag{6.59}$$

$$P_{jk} = \text{Prob}\,(u_{jk} > u_{j'k'}, j', k' \neq j, k) \tag{6.60}$$

$$= \int_{0}^{\infty} du\, \Gamma_{jk}(u, u_{jk}^*)\theta_{jk}(u) \tag{6.61}$$

$$= \int_{0}^{\infty} du\, \Gamma_{jk}(u, u_{jk}^*) \prod_{j'k' \neq jk} \int_{-\infty}^{u} du' \Gamma_{j'k'}(u', u_{j'k'}^*) \tag{6.62}$$

(if we substitute for θ_{jk} from (6.59) into (6.67)). To evaluate this expression, an assumption is needed about the distribution of ε_{jk}. This is usually taken as a Weibull distribution, which has mean zero and standard deviation σ given by

$$\sigma = \frac{\Pi}{\sqrt{6\lambda}}. \tag{6.63}$$

λ is the parameter of the distribution which can be written as

$$\Gamma_{jk}(\varepsilon_{jk}, \lambda) = \Gamma(u - u_{jk}^*, \lambda) \tag{6.64}$$

$$= \lambda e^{-\lambda \varepsilon_{jk}} e^{-e^{-\lambda \varepsilon_{jk}}} \tag{6.65}$$

Note that the notation Γ_{jk} can be used, as (6.64) shows, because the probability that ε_{jk} takes a particular value is the same as the probability that u takes a particular value. The scales are shifted by the mean u_{jk}^* but since the range of integration of u in equation (6.61) is from 0 to ∞, this is of no consequence. Thus, if Γ_{jk} is substituted from (6.65) into (6.62), P_{jk} can be calculated. Though this looks difficult it can be done and the answer takes the multinomial logit form

$$P_{jk} = \frac{e^{\lambda \mu_{jk}^*}}{\sum_{jk} e^{\lambda \mu_{jk}^*}} \tag{6.66}$$

(which is a form of equation (6.56)), where λ is the parameter of the distribution. If we make the simplest possible assumption for u_{jk}, of an additive utility function (with components z_q^{jk} where coefficients are then the parameters, α), then

$$u_{jk} = \sum_{q} \alpha_q z_q^{jk}. \tag{6.67}$$

P_{jk} can be written

$$P_{jk} = \frac{e^{\lambda \sum_q \alpha_q z_q^{jk}}}{\sum\limits_{jk} e^{\lambda \sum_q \alpha_q z_q^{jk}}}. \tag{6.68}$$

This derivation of P_{jk} assumes that ε_{jk} in equation (6.58) itself is Weibull distributed. The situation is not always so clear when we look at the individual components of utility. A more general form of utility function relates to the three kinds of attributes which were identified at the outset — j-dependent, k-dependent, and (j, k)-dependent. In such a case, we can write

$$u_{jk} = u_j^{(1)*} + u_k^{(2)*} + u_{jk}^{(3)*} + \varepsilon_j^{(1)} + \varepsilon_k^{(2)} + \varepsilon_{jk}^{(3)}. \tag{6.69}$$

Three random variables are involved and they need not combine and behave like a single Weibull-distributed random variable ε_{jk}. Indeed, the general choice model associated with equation (6.69) can only be obtained in an approximate form. First, then, consider the slightly simpler case where we assume no pure k variation. (For the journey to work, this is equivalent to assuming components of utility dependent on the job alone (such as wage) and on fare to that location by a particular mode (such as cost of travel), but not on mode alone.)

$$u_{jk} = u_j^* + u_{jk}^* + \varepsilon_j + \varepsilon_{jk} \tag{6.70}$$

(where we now drop the numerical superscripts because it is no longer necessary to distinguish u_j^* from u_k^* for numerical values of j and k). This generates a model which can be written sequentially as the product of two logit formulae as

$$P_{jk} = P_j P_{k|j}$$

$$= \frac{e^{\beta u_{j*}^{+*}}}{\sum\limits_j e^{\beta u_{j*}^{+*}}} \cdot \frac{e^{\lambda u_{jk}^*}}{\sum\limits_k e^{\lambda u_{jk}^*}} \tag{6.71}$$

where

$$u_{j*}^{+*} = u_j^* + \tilde{u}_{j*} \tag{6.72}$$

where $\tilde{u}_{j*}$ is defined by

$$e^{\lambda \tilde{u}_{j*}} = \sum\limits_k e^{\lambda u_{jk}^*} \tag{6.73}$$

which is

$$\tilde{u}_{j*} = \frac{1}{\lambda} \log \sum\limits_k e^{\lambda u_{jk}^*}. \tag{6.74}$$

This implies that the decision process can be viewed as a sequential one (at least as a piece of mathematics) as: (a) select j to maximize $u_j^* + \tilde{u}_{j*}$; (b) given j, choose k to maximize u_{jk}. This only works, however, if the quantity $\tilde{u}_{j*}$ is added to u_j^* at the j-level. This is a *composite utility* which is a measure of the k-average —u_{jk} utility

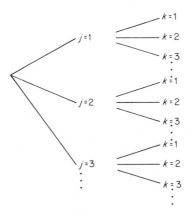

Figure 6.3 Tree structure

associated with u. It is a model construct and must satisfy equation (6.74) (or, equivalently, (6.73), which is easier to remember).

The standard deviation of u_{j*}^{+*} in P_j in equation (6.71) is

$$\sigma_{j*}^{+} = \sqrt{\sigma_j^2 + \tilde{\sigma}_{j*}^2} \tag{6.75}$$

$$= \sqrt{\sigma_j^2 + \pi/6\lambda} \tag{6.76}$$

and since

$$\sigma_{j*}^{+2} = \frac{\pi}{6\beta} \tag{6.77}$$

these equations imply

$$\beta \leqslant \lambda \tag{6.78}$$

which is an important result because it offers a theoretical relationship constraining parameter size.

The utility function given by equation (6.70) governs a tree-like decision process as indicated in Figure 6.3. This argument can easily be extended to cover more steps in the tree.

The main implications of random utility theory for modelling are drawn in the next subsection since it is more interesting in the context of this book to do this in comparison with entropy maximizing modelling.

6.2.7 Programming, random utility, and entropy maximizing

The first point of remark is an obvious one in the light of Section 6.2.5: entropy maximizing models are special cases of nonlinear programming models which arise from problems which can be solved by Lagrangian methods. One useful consequence of this is that use can be made of the dual. Equation (6.52), for example, gives T_{ij} in terms of dual variables:

$$T_{ij} = e^{-\alpha_i - \nu_j - \beta c_{ij}} \tag{6.79}$$

and if this is substituted into the Lagrangian (6.51) then the problem becomes an unconstrained one in the dual variables. There are standard techniques for solving this and because there are far fewer dual variables than primal variables ($2N + 1$ against N^2), this is beginning to be a useful computational procedure, especially for calibration.

It also turns out that there is a close relationship between the *linear transportation problem* of linear programming and the particular *nonlinear* entropy maximizing model which involves the same primal variables and mostly the same constraints. The main equations are restated here for convenience. The primal linear programming problem is

$$\text{Min}_{\{T_{ij}\}} Z = \sum_{ij} T_{ij} c_{ij} \tag{6.80}$$

such that

$$\sum_j T_{ij} = O_i \tag{6.81}$$

$$\sum_i T_{ij} = D_j. \tag{6.82}$$

The dual is

$$\text{Max}_{\{\alpha_i, v_j\}} Z' = \sum_i \alpha_i O_i + \sum_j v_j D_j \tag{6.83}$$

such that

$$c_{ij} \geqslant \alpha_i + v_j. \tag{6.84}$$

The entropy maximizing model is

$$\text{Max}_{\{T_{ij}\}} S = - \sum_{ij} \log T_{ij}! \tag{6.85}$$

such that (6.81) and (6.82) hold as above *together with*

$$\sum_{ij} T_{ij} c_{ij} = C \tag{6.86}$$

where C is to be identified with Z for present purposes. The dual is

$$\text{Min}_{\{\lambda_i^{(1)} \lambda_j^{(2)} \beta\}} S' = \sum_i \lambda_i^{(1)} O_i + \sum_j \lambda_j^{(2)} D_j + \beta C \tag{6.87}$$

such that

$$c_{ij} = \frac{\lambda_i^{(1)}}{\beta} - \frac{\lambda_j^{(2)}}{\beta} - \log \frac{T_{ij}}{\beta}. \tag{6.88}$$

The dual constraints, (6.88), can be rearranged and solved for T_{ij} as

$$T_{ij} = e^{-\lambda_i^{(1)}} e^{-\lambda_j^{(2)}} e^{-\beta c_{ij}} \tag{6.89}$$

$$= A_i O_i B_j D_j e^{-\beta c_{ij}} \tag{6.90}$$

where

$$e^{-\lambda_i^{(1)}} = A_i O_i \qquad (6.91)$$

$$e^{-\lambda_j^{(2)}} = B_j D_j. \qquad (6.92)$$

The last three equations show that, with these definitions for the balancing factors in terms of multipliers, the dual constraints are the main model equations usually derived from the primal.

There is a $(1-1)$ relation between β in the EM model equations and C which appears in the third constraint (6.86). As β increases, C decreases, as indicated on the plot in Figure 6.4 of C. It seems intuitively clear that, as indicated on the figure, there is a minimum value C_{min} for C and that the curve will be asymptotic to $C = C_{min}$ as β becomes very large (written $\beta \to \infty$). However, C_{min} is, or should be, the solution of the linear programming problem. Thus,

$$C(EM) \to C(LP) = C_{min} \text{ as } \beta \to \infty \quad (\text{and } C_{min} = Z). \qquad (6.93)$$

It can also be shown that

$$T_{ij}(EM) \to T_{ij}(LP) \text{ as } \beta \to \infty. \qquad (6.94)$$

There are also relationships between the dual variables:

$$\frac{-\lambda_i^{(1)}}{\beta} \to \alpha_i \text{ as } \beta \to \infty \qquad (6.95)$$

$$\frac{-\lambda_j^{(2)}}{\beta} \to v_j \text{ as } \beta \to \infty. \qquad (6.96)$$

If we recall that α_i and v_j were measures of comparative advantage, it turns out that in the EM case, the comparative advantage represented by the dual variables is less (as one would expect because in a sense the EM model is suboptimal).

It remains to point out the contrast between the linear programming and entropy maximizing modelling styles. The main contrast can be seen intuitively if we recall that in the LP case there are relatively few nonzero values of the T_{ij}-variables ($2N + 1$ at most as distinct from N^2 in the EM case). Real data sets are

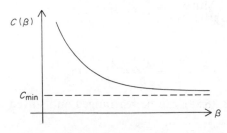

Figure 6.4 Travel cost and $\beta \to \infty$ limit

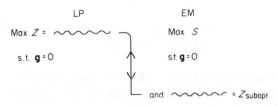

Figure 6.5 LP–EM relationships

usually more like the EM case with large numbers of nonzero variables. This is because people, in aggregate, are unlikely to be optimizing. There will be market imperfection, lack of information on alternatives, and so on, which will ensure that the overall distribution in suboptimal.

The main relationship between the two styles is shown in Figure 6.5. The LP model can be converted to a corresponding EM model by taking the objective function and converting it to be a *constraint* but taking a suboptimal value; or vice versa.

We can now draw together several threads of argument and compare random utility, entropy maximizing, and other models arising from optimizing procedures. As a first step, we need to get the models in a form in which they are most easily comparable. Consider the disaggregated entropy maximizing model given as equation (4.43) of Chapter 4:

$$T_{ij}^{kn} = A_i^n B_j O_i^n D_j e^{-\beta^n c_{ij}^k}. \tag{6.97}$$

This can be generalized slightly by noting that the parameters β^n are made to do two jobs: to measure sensitivity to destination choice (c_{ij}^k variation with j) and to model choice (c_{ij}^k variation with k). We can introduce two sets of parameters, β^n and, say, λ^n, to cope with each of these in turn. $e^{-\beta^n c_{ij}^k}$ is then replaced by

$$e^{-\beta^n c_{ij}^k} = e^{-\beta^n C_{ij}^n} \frac{e^{-\lambda^n c_{ij}^k}}{\sum_k e^{-\lambda^n c_{ij}^k}} \tag{6.98}$$

$$= e^{-\beta^n C_{ij}^n} M_{ij}^{kn} \tag{6.99}$$

where

$$M_{ij}^{kn} = \frac{e^{-\lambda^n c_{ij}^k}}{\sum_k e^{-\lambda^n c_{ij}^k}} \tag{6.100}$$

is the proportion of n-type (i,j) travellers using mode k. λ^n now measures sensitivity to modal choice. β^n to destination choice. C_{ij}^n is a measure of travel cost from i to j which takes account of cost by available modes, k, but in some 'average' way—it is itself not dependent on k. It is known as a composite cost and is some function of the c_{ij}^k's. As we shall see later, we take

$$e^{-\lambda^n C_{ij}^n} = \sum_k e^{-\lambda^n c_{ij}^k} \tag{6.101}$$

which is,

$$C_{ij}^n = \frac{1}{\lambda^n} \log \sum_k e^{-\lambda^n c_{ij}^k} \tag{6.102}$$

(a kind of 'exponential averaging'). This is simply the 'cost' version of the composite utility derived as equation (6.74) above. The model given by equation (6.97) can now be modified to

$$T_{ij}^{kn} = A_i^n B_j O_i^n D_j e^{-\beta^n C_{ij}^n} M_{ij}^{kn}. \tag{6.103}$$

M_{ij}^{kn} is known as the *modal share* of mode k, and we now seek to write the whole model in 'share' form. Let M_{ij}^n be the share of (i, n) trips to destination j. Then, (6.103) can be written

$$T_{ij}^{kn} = O_i^n M_{ij}^n M_{ij}^{kn} \tag{6.104}$$

with M_{ij}^{kn} given by (6.100) and

$$M_{ij}^n = \frac{B_j D_j e^{-\beta^n C_{ij}^n}}{\sum_j B_j D_j e^{-\beta^n C_{ij}^n}} \tag{6.105}$$

since A_i^n is

$$A_i^n = \frac{1}{\sum_j B_j D_j e^{-\beta^n C_{ij}^n}}. \tag{6.106}$$

For later convenience, we can recall

$$B_j D_j = e^{-\lambda_j^{(2)}} \tag{6.107}$$

from equation (6.92) and write equation (6.105) as

$$M_{ij}^n = \frac{e^{-\lambda_j^{(2)} - \beta^n C_{ij}^n}}{\sum_j e^{-\lambda_j^{(2)} - \beta^n C_{ij}^n}}. \tag{6.108}$$

Thus equations (6.104), (6.108) (or (6.105)), and (6.100) in sequence constitute a formulation of the disaggregated entropy maximizing model in general share form.

The random utility method predicts shares, too, as in equation (6.71). We now add the i-subscripts and n-superscripts, where appropriate, and write

$$u_{jk}^* = -c_{ij}^k \tag{6.109}$$

$$\tilde{u}_{j*} = -C_{ij}^n = -\frac{1}{\lambda^n} \log \sum_k e^{-\lambda^n c_{ij}^k} \tag{6.110}$$

$$u_j^* = U_{ij}^n + b_j \tag{6.111}$$

where U_{ij}^n is the positive utility associated with the (i, j) choice for type n people and b_j is a set of 'prices' associated with any j-constraints. Then,

$$u_{j*}^{+*} = U_{ij}^n + b_j - C_{ij}^n \tag{6.112}$$

and equation (6.71) becomes

$$P_{jk}(i,n) = \frac{e^{\beta^n(b_j + U_{ij}^n - C_{ij}^n)}}{\sum\limits_j e^{\beta^n(b_j + U_{ij}^n - C_{ij}^n)}} \cdot \frac{e^{-\lambda^n c_{ij}^k}}{\sum\limits_k e^{-\lambda^n c_{ij}^k}} \tag{6.113}$$

with C_{ij}^n given in terms of c_{ij}^k's by (6.110). We have added (i,n) to P_{jk} to show that this is now the probability for a type-n resident of zone i. Thus

$$T_{ij}^{kn} = P_{jk}(i,n)O_i^n = O_i^n M_{ij}^n M_{ij}^{kn} \tag{6.114}$$

where

$$M_{ij}^n = \frac{e^{\beta^n(b_j + U_{ij}^n - C_{ij}^n)}}{\sum\limits_j e^{\beta^n(b_i + U_{ij}^n - C_{ij}^n)}} \tag{6.115}$$

and M_{ij}^{kn} as in equation (6.100). So, the equations to be compared are (6.108) and (6.115)—the rest of the structure being formally identical.

Two points stand out, both of which will be taken up again shortly: first, there is an additional utility term U_{ij}^n; secondly, the term b_j which plays the same role as $\lambda_j^{(2)}$ in ensuring that the D_j constraints are satisfied is multiplied by β^n here whereas $\lambda_j^{(2)}$ was not in equation (6.108). b_j has dimension of utility, or cost, while $\lambda_j^{(2)}$ is dimensionless.

Now that the differences between the entropy maximizing and utility maximizing models (*when presented at the same level of resolution*) have been described and clarified, a number of points can be made.

(1) The random utility approach is rooted in the micro scale and has a direct behavioural interpretation. The entropy maximizing approach is *essentially* meso scale, as explained in Chapter 4, though can be related to the micro scale through probabilities and information theory as discussed in Section 4.8.

(2) One immediate practical point follows from this discussion. In entropy maximizing models, c_{ij}^k is usually taken as a generalized cost (as in equation (4.60) with k and n superscripts added as appropriate):

$$c_{ij}^k = a_1^n t_{ij}^k + a_2^n d_{ij}^k + a_3^n e_{ij}^k + \ldots + p_j^k \tag{6.116}$$

(where we take the last term, with a unit coefficient, as terminal charges for mode k at j). The coefficients $a_1^k, a_2^k, a_3^k, \ldots$ are usually taken as given. However, if a set of person data is available, then data on M_{ij}^{kn} and a version of equation (6.100) can be used to estimate these.

$$M_{ij}^{kn} = \frac{e^{-\lambda^n(a_1^n t_{ij}^k + a_2^n d_{ij}^k + a_3^n e_{ij}^k + \ldots + p_j^k)}}{\sum\limits_k e^{-\lambda^n(a_1^{kn} t_{ij}^k + a_2^{kn} d_{ij}^k + a_3^{kn} e_{ij}^k + p_j^k)}} \cdot \tag{6.117}$$

If we write out (6.117) again for k' and divide the two equations and take logs, we get

$$\log\left(\frac{M_{ij}^{kn}}{M_{ij}^{k'n}}\right) = -\lambda^n a_1^n(t_{ij}^k - t_{ij}^{k'}) - \lambda^n a_2^n(d_{ij}^k - d_{ij}^{k'}) - \lambda^n a_3^n(e_{ij}^k - e_{ij}^{k'})$$

$$-\cdots - \lambda^n(p_j^k - p_j^{k'}) \tag{6.118}$$

$$= b_1^n(t_{ij}^k - t_{ij}^{k'}) + b_2^n(d_{ij}^k - d_{ij}^{k'}) + b_3^n(e_{ij}^k - e_{ij}^{k'})$$

$$-\lambda^n(p_j^k - p_j^{k'}) \tag{6.119}$$

where

$$b_1^n = -\lambda^n a_1^n, b_2^n = -\lambda^n a_3^n, b_3^n = -\lambda^n a_3^n, \ldots . \tag{6.120}$$

The coefficients $b_1^n, b_2^n, b_3^n, \ldots, \lambda^n$ can now be estimated from (6.119) by regression analysis and $a_1^{kn}, a_2^{kn}, a_3^{kn}, \ldots$ calculated from (6.120). A coefficient such as a_1^{kn} is then the money value-of-time per type n people.

(3) In setting up the entropy maximizing model in share form, equation (6.101) (or (6.102)) has to be assumed as the appropriate functional form for composite cost (though it was recognized that a degree of internal consistency was achieved by this). In random utility theory, this emerges naturally as the only form, though the derivation of this would take us beyond our present scope.

(4) Equation (6.78) implies that, for this disaggregated case, we should have

$$\beta^n < \lambda^n. \tag{6.121}$$

In practice, this often does not hold. A possible explanation of this is that the U_{ij}^n term is omitted from the utility function in transport applications. This suggests that, were $\beta^n > \lambda^n$ when the model (6.103) is calibrated, it should be rewritten as

$$T_{ij}^{kn} = A_i^n B_j O_i^n D_j e^{(\beta^{n'} - \beta^n)C_{ij}^n} e^{-\beta^{n'}C_{ij}^n} M_{ij}^{kn} \tag{6.122}$$

where β^n is the calibrated value and $\beta^{n'}$ is *chosen* so that (6.121) holds. Then $e^{(\beta^{n'} - \beta^n)C_{ij}^n}$ can be taken as an estimate of $e^{\beta^{n'}U_{ij}^n}$:

$$e^{\beta^{n'}U_{ij}^n} = e^{(\beta^{n'} - \beta^n)C_{ij}^n} \tag{6.123}$$

and can be used as such for forecasting.

(5) Finally, we comment on the difference between the $\lambda_j^{(2)}$ and b_i terms. $\lambda_j^{(2)}$ arises because the entropy-maximizing objective function takes the form

$$-\sum_{ijkn} T_{ij}^{kn} \log T_{ij}^{kn}. \tag{6.124}$$

b_j arises because the objective function *which would produce the aggregated random utility model* is

$$-\sum_n \frac{1}{\beta^n} \sum_{ijk} T_{ij}^{kn} \log T_{ij}^{kn}. \tag{6.125}$$

In other words, each entropy term is *weighted* by $1/\beta^n$. This second objective function is known as the *group surplus* function because of its connection to the consumers' surplus implied by the aggregated random utility demand function.

In the light of these comparisons, what can be said in conclusion about entropy maximizing in relation to utility maximizing? The approaches do involve different identifying assumptions and different levels of resolution and so the basic issue for the model builder is: what set of assumptions should be used for particular circumstances at a particular time? If the entropy maximizing method is then chosen, however, the random utility method at the very least offers a composite cost formula together with strong advice on the relationship between parameters such as β^n and λ^n.

6.3 CONCLUDING COMMENTS

Some of the main methods for optimization have been outlined in this chapter. In this concluding section, we summarize the main features of the examples used and note what has been omitted. We also note some links to other treatments and examples of optimization in subsequent chapters.

We noted at the outset of this chapter that the term 'optimization' is used in a technical sense as being concerned with the minimization or maximization of functions, possibly subject to constraints. The associated methematics can then be used either for building analytical models of phenomena or behaviour, or in a planning context. The examples used to illustrate the mathematics above are almost wholly concerned with analytical and behavioural models. This has also led to a predominant concern with urban systems as the source of illustration. Planning applications will be illustrated in the context of planning and control theory in Chapters 9 and 10.

The methods outlined here are applied to models which are essentially static. If models based on such methods are used for forecasting, then this involves 'comparative static' assumptions: in effect, if a parameter or other variable changes, then the system moves quickly to the new optimum. Thus, transient effects can be ignored. The optimum is usually some kind of 'equilibrium' and in this sense the notion ties up with one of the most important basic concepts of Chapter 2.

We will see later that 'comparative static' change need not be smooth, and this makes it more interesting than is often assumed. Indeed, we have already had an indication of this in the earlier remark that the solution to a linear programming problem is at a 'corner' of the polyhedron formed by the constraints. When one or more parameters change, the position of the corners will change. If the optimum remains at the same corner, then the transition will be smooth. But if the optimum is at a different corner after the change, a discrete jump in the system state will be involved. We set this kind of change in a much broader context in the discussion of catastrophe theory in Chapter 8.

The result outlined in the previous paragraph arose out of the notion of the 'constraint' set which is obviously important to all kinds of mathematical programming. We have seen above that the constraints associated with the entropy-maximizing models of Chapter 4 can be taken on those of a nonlinear programming problem. Also, many of the accounting equations derived in Chapter 5 can be considered as constraint equations: they relate the system's variables at all points in time. The general point to be noted is that much of the interdependence relating the variables of a system can be represented in constraint equations and built into models using the methods of this or earlier chapters. It is useful to observe this as a general point.

The main omission of this chapter relates to more general methods for treating optimization over time. These come into play when parameters of a model are changing in some complicated way and when there are other parameters, the control variables, where values are to be set optimally over a period of time. The constraints can now include differential equations — a notion which will be taken further in Chapter 8. There are two main classes of method for these problems: control theory and dynamic programming. However, they will not be pursued here. There have been relatively few applications in environmental studies, mainly because although the methods are theoretically elegant, it is difficult in practice to apply them to problems where there is a large number of variables — and this, of course, is the situation which is typical of systems analysis. Some appropriate references are given in the notes on further reading below for readers who want to take their study of these methods further.

Finally, we note that there are some specific and complicated optimization concepts associated with networks. This partly relates to network equilibrium and partly to mechanisms for network development. Both these topics are dealt with in the broader context of network analysis in Chapter 7.

NOTES ON FURTHER READING

There is a huge mass of literature on mathematical programming and optimization methods. A fairly brief general introduction is provided by †Wilson, Coelho, Macgill, and Williams (1981, Chapter 2). The *rest of that book provides a wide range of examples of optimization methodology applied to transport and locational analysis. Random utility theory is reviewed in the same book (*Chapter 4) and in *Williams (1977) (6.2.6). The relationship between entropy-maximizing doubly-constrained spatial-interaction models and the transportation problem of linear programming was first proved by *Evans (1973). It was extended more widely, with some results on duals added, by *Wilson and Senior (1974) (6.2.7). Dynamic programming is described briefly by *Miller (1979) and a control theory example, based on the work of Ptrin, by *Wilson (1981).

CHAPTER 7

Locational and network structures: nodes and links

7.1 NODES AND LINKS AS SYSTEM COMPONENTS

7.1.1 The nature of network analysis

Networks are made up of nodes and links, as in the example in Figure 7.1. Because the nodes may carry activities, or have capacities which are significant for the performance of the network, and because such nodes are located (at least relative to each other in space), it is convenient to broaden the usual concept of 'network structure' to 'locational and network structures'. The main purpose of this chapter is then to explore aspects of systems analytical methods which are peculiar to networks in this sense.

Networks in environmental systems carry *flows* of some kind between locations. Nodes at which flows arise are called *sources*; those at which flows terminate are called *sinks*; other nodes are intermediate. For example, in a representation of a journey to work, a network may carry a flow from a group of homes (the source) to a group of jobs (the sink) on road links via intermediate nodes which are junctions on the network. Such a flow is illustrated in Figure 7.2 from a source A to a sink B via intermediate nodes X, Y, Z.

Perhaps the main feature of network analysis which demands methods which differ in principle from those we have seen in earlier chapters is that links share flows. Figure 7.3 is similar to Figure 7.2, but a new source, C, has been added, together with a new flow from C to B. Note that the links XY, YZ, and ZB now

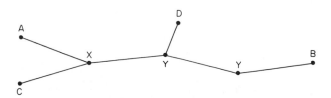

Figure 7.1 An example of a network

133

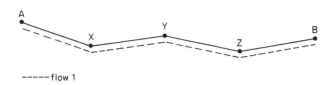

Figure 7.2 A network route with flows

each carry two flows. In general, it is easy to see that any total network *link flow* is made up of a set of *source–sink* (or *origin–destination*) flows.

In earlier studies of flow, we worked directly with origin–destination flows, say as an array $\{T_{ij}\}$, where the sets of i's and j's are sources and sinks respectively and we ignored the network links, or channels, which carry such flows, as exhibited in Figure 7.3. Other system properties, such as the cost of travel between i and j, c_{ij}, were implicitly assumed to be independent of the underlying network. In fact, this is not the case, and we need new methods of model building to patch up a deficiency here. The essence of the problem can be seen through a deeper investigation of the notion of generalized cost introduced in Chapter 4 and equation (4.60). Let us assume it takes the form

$$c_{ij} = a_1 P_i^{(O)} + a_2 t_{ij} + a_3 d_{ij} + a_4 P_j^{(D)} \tag{7.1}$$

where $P_i^{(O)}$ and $P_j^{(D)}$ are origin and destination nodal, or terminal, costs (say due to 'waiting time' or 'parking charges'), $a_2 t_{ij}$ is a term proportional to travel time, and $a_3 d_{ij}$ a term proportional to distance (say fuel costs). Let us now focus on the t_{ij} term: how would it actually be measured? In Figure 7.2 (or 7.3), we can see that it is a sum of travel times on links — using an obvious notation, it is

$$t_{ij} = t_{AX} + t_{XY} + t_{YZ} + T_{ZB} \tag{7.2}$$

(if $i =$ A and $j =$ B) in that particular example. But each link time, say t_{XY}, will be a function of congestion on that link and *this is dependent on all the flows which share the link*. Thus, to get one t_{ij} term, information will be needed on all flows and on how they are '*assigned*' to the network. Since the flows are themselves dependent on c_{ij} (and hence t_{ij}), we now perceive a problem of mutual interdependence — T_{ij}

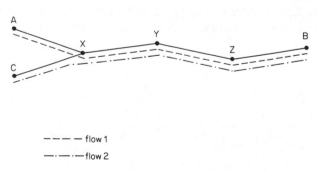

Figure 7.3 Two network routes with flows

on c_{ij}, c_{ij} on *all* the T_{ij}'s—which looks difficult to sort out. We show how to resolve this problem in Section 7.2, but this illustrates why techniques of network analysis are needed. Further, the problem could be made even more complicated if terms such as $P_i^{(O)}$ and $P_j^{(D)}$ were dependent on traffic and capacities at nodes, and if t_{ij} was dependent on traffic at intermediate nodes—as in fact it does depend on capacity at intermediate road junctions in the traffic case.

7.1.2 Nodes and links in systems

We have seen that couplings between components play an important role in determining the nature of behaviour in particular systems. Such couplings are usually flows of some kind carried on links between nodes, and so we can expect network structures to be of the greatest importance. In this subsection, we consider some of the most obvious nodal and link structural features of the three kinds of system considered as examples in Part 1.

In moorland ecosystems, the most obviously visible networks are at the biological rather than the ecological (or geographical) scale: they are the (literally) tree-like structures of the plants themselves and their root systems. These structures provide the channels for the flow of water and nutrients in the plants. The soil itself should also be regarded as a network link in the various cycles, and the pattern of drainage channels on the surface may be significant as well.

For water-resource systems, the distribution network is one of the most obvious and visible features. Sources are particular drainage basins or reservoirs, and sinks are the extraction points from the network for consumption. Such networks have already been exhibited in Chapter 1—as in Figure 1.8 for example. In practice, water networks are linked with other networks—particularly those for sewage and other waste disposal.

Cities provide the richest of the three examples in this case. As well as the physically obvious transport networks (of various modes), there are many others: telephone, mail, radio, TV, sewage, water, gas, electricity, hot water/steam, computers, and so on. Many of these are interdependent. These networks exhibit different kinds of characteristics. Two transport modes, for example, bus and private car, share the road network, whereas the rail network is a single-mode physical structure. An air network on the other hand is focused more on nodes and vehicles rather than network links (though in heavily trafficked air space, 'corridors' will be used by air traffic controllers which play the role of links). Radio and TV networks are also examples of nodes (transmitters and receivers) being more important than network links. Sometimes intermediate nodes play a major functional role: for example, switching nodes (exchanges) in telephone systems.

7.1.3 Types of network: trees and circuits

There are two main types of network evident in the examples cited in the preceding subsection. Plants exhibit, literally, *tree*-like structures, and a little

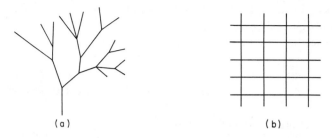

Figure 7.4 (a) Tree network; (b) network with circuits

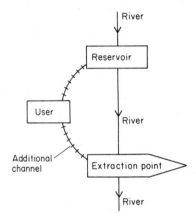

Figure 7.5 A natural tree network
with added links

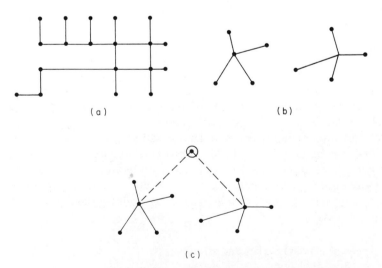

Figure 7.6 (a) A fully connected network; (b) a not-fully connected network; (c) central nodes connected by higher level of organization

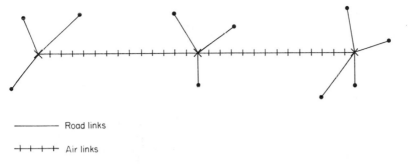

——————— Road links

+ + + + + Air links

Figure 7.7 A multi-modal fully-connected network

thought shows that the distinguishing characteristic is that *circuits* are not possible in such networks. Examples of the two kinds are shown in Figure 7.4. It is also interesting that the cyclic processes in ecosystems illustrate flows round networks which are *essentially* circuits.

Natural river systems are also tree-like structures. This arises because of the nesting and interlocking of drainage basins within a hierarchy (of which concept, more later). Such networks may be modified to include some circuits when additional channels are added, as in the example in Figure 7.5.

Most urban networks are grid-like networks in which circuits are possible. We will see shortly, however, that transport networks can be seen as overlapping superimposed tree networks. This is because the network made up of the best routes from a single node to all other nodes is a tree and so, if there are N nodes, a transport network can be said to be made up of N such trees.

It is also perhaps worth noting the distinction between networks whose nodes are wholly connected and those where this is not true. Nodes are connected if a route, perhaps involving intermediate nodes, exists between them. Examples are shown in Figure 7.6. The example shown in Figure 7.6(b) could, for obvious reasons, be called a 'star system' and these often occur when sources are assigned, say, to the 'nearest' sink — as may be the case with young children from homes to primary schools in cities. In such cases, a higher level of organization may connect the 'central' nodes as shown in Figure 7.6(c) — which may be the city education office 'planning' the allocation of children to schools.

This last example begins to illustrate the mutual interdependence of several networks. Another obvious example arises in the transport field: small towns without airports may be disconnected from the airline network if such a network is mapped directly. However, if it is shown coupled to a road network, as happens in practice, then such systems become wholly connected, as shown in Figure 7.7.

7.1.4 More abstract networks in systems

So far, we have concentrated on easily visible network structures. Network analysis can also be helpful in relation to more abstract structures which can be identified in systems, and some examples of such ideas are sketched briefly here.

138

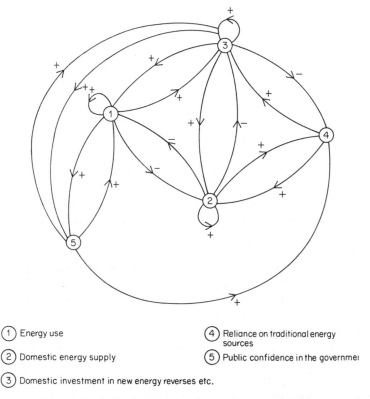

(1) Energy use

(2) Domestic energy supply

(3) Domestic investment in new energy reverses etc.

(4) Reliance on traditional energy sources

(5) Public confidence in the governmei

Figure 7.8 A signed digraph (after Roberts, 1976, p196)

First, in some abstract space, system components can be taken as nodes, and such nodes joined by links (with an arrow indicating the direction of causality) if change in one implies change in another. If the change is known as positive or negative, then an appropriate sign can be associated with each link. Such network representations of system interdependence are known as directed graphs, or *digraphs* for short; if signs are added, they are known as *signed digraphs*. These graphs are typically circuits, of course, since they will include the various feedbacks of the system. An example is shown in Figure 7.8. Such graphs have

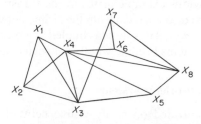

Figure 7.9 *Q*-structures (after Atkin, 1977, p16, Figure 1)

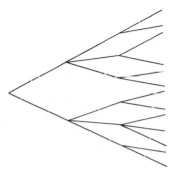

Figure 7.10   Decision trees

obvious connections with the more general notion of system diagrams introduced in Chapter 2, and help to provide the basis for Forrester's treatment of system dynamics in the next chapter.

A second kind of abstract structure arises out of the work of Atkin and his colleagues: so-called q-analysis. This involves the grouping of system components into sets of components called complexes and then analysing structural connections between them. The basic idea is displayed in Figure 7.9, and the technique will be taken up again in more detail below.

The next kind of abstract graph arises when uncertainty or planning is a process within the system and decisions can often be represented in a hierarchical tree-like network as shown in Figure 7.10. Each branching node represents a new decision point and the links extending below it represent alternatives. The analysis of such graphs, especially the use of so-called branch-and-bound methods, is an important tool of network analysis and we return to it again below.

7.2 SOME TECHNIQUES OF NETWORK ANALYSIS

7.2.1 Network description

A network consists of a set of nodes and a set of links. Each link is identified by the node numbers at either end (in beginning–end order if a direction is implied). Thus, the essence of network description is the numbering of nodes, as in Figure 7.11. In that example, there are 11 nodes, and the boxed ones, 1, 3, 4, 5, 10, and 11 can be sources and sinks – the others are intermediate nodes. There are 17 links: (1, 2), (1, 6), (1, 7), (2, 6), (6, 7), (2, 3), (6, 3), (6, 8), (3, 4), (3, 5), (4, 5), (3, 8), (7, 8), (8, 11), (8, 9), (9, 10), (11, 10). Since no directions are identified, symmetry is assumed so that travel (or whatever) on (1, 2) is assumed to have the same characteristics as on (2, 1). If symmetry cannot be assumed — as is common in traffic networks — then two-way links have to be identified, as in Figure 7.12, making 34 in all.

In order to develop a suitable notation further, it is useful to use the conventions of set theory. For example, S may be the set of nodes which can be

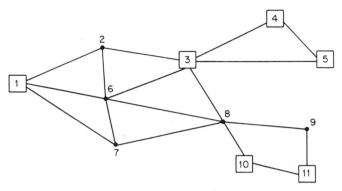

Figure 7.11 Example of a network, with sources and sinks

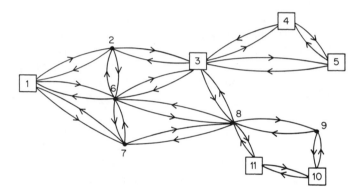

Figure 7.12 Figure 7.11, with two one-way links between each
connected pair of nodes

sources and sinks, and $i \in S$ denotes a node in that set*. S is the set $(1, 3, 4, 5, 10, 11)$.
Thus, in the traffic case, a flow matrix $\{T_{ij}\}$ would be defined for $i, j \in S$.

The definition and construction of a network is analogous in an important
respect to the task of defining a zoning system: it has to be appropriate for the
particular analytical task in hand and a suitable level of resolution has to be
adopted. Traffic modelling provides an example for discussion, and similar points
can be made for other networks. If each junction in a traffic network is taken as a
node of the model network, then this will usually correspond to a rather fine level
of resolution. This, however, may be the appropriate scale for analysis.
Alternatively, all junctions of some set of major roads only may be taken as the
nodes of a network.

Once the broad decision has been taken on which roads and junctions to
include, a finer level of resolution may be reached by adding more detail near
important nodes—particularly sources and sinks. Suppose, for example, that

*Recall that '∈' is the set inclusion sign.

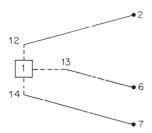

Figure 7.13 Addition of
dummy links

Figure 7.11 is the first attempt at representing a network. Then node 1 may be the centroid of a housing estate and the three links connected to it may be different roads into it. A more realistic representation of the network near the node could then be taken as that shown in Figure 7.13. Three notional links (1, 12), (1, 13), and (1, 14) have been added which represent, say, the *different* average travel times to get from the centre of zone 1 to the three main roads out of the zone. Addition of such detail quickly adds a lot of links to the network description of course.

Alternatively, it may be useful in a particular study to move to a coarser scale and an example of this is a *spider* network. Here, notional links join the main nodes — probably the sources and sinks — only. Figure 7.11, for example, may then be condensed to something like Figure 7.14.

The next step in building up an effective description of a network is to list the properties of interest associated with links or nodes. We saw in equations (7.1) and (7.2), for example, that a number of elements contributed to generalized cost, c_{ij} (7.1), and that components such as travel time were made up of sums of link elements (as in equation (7.2)). Thus, we need variables to describe such link properties. Some link properties may be functionally related to each other. For example, travel time (or speed) on a link will depend on the traffic flow on that link. Indeed, it is the existence of such relationships that add difficulty (and

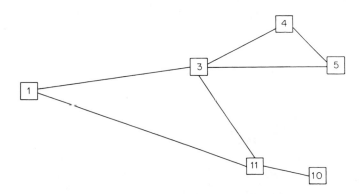

Figure 7.14 A spider network

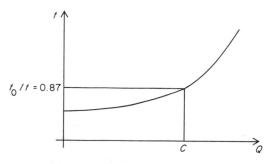

Figure 7.15 Time-flow relationship

interest !), in the form of nonlinearities, to a number of problems. We therefore explore examples of such relationships before proceeding.

Let Q be a flow on some link, t the travel time, and s the speed. Of course, $t = d/s$, where d is the link length. Many possible functional relationships have been used, but two commonly used ones will be mentioned here. In America, it is common to relate travel time to flow by the formula

$$t = t_0 \left[1 + 0.15 \left(\frac{Q}{C} \right)^4 \right] \tag{7.3}$$

where C is a constant. t_0 is another constant which gives the free-flow time. This relationship is graphed in Figure 7.15. The constant C is the *design capacity* for that link: the flow which can be carried which gives $t_0/t = 0.87$. Note that such a definition involves a degree of arbitrariness (as we saw with 'accessibility' in an earlier context).

In Britain, speed is related to flow as shown in Figure 7.16. Speed is assumed to take a free flow value up to some value of Q. The design capacity in this case would be the flow C which corresponded to some design speed s_D, as shown on the figure.

These time-flow or speed-flow relationships will be used in the discussion of shortest paths and assignment of flows to networks in the next subsection. Meanwhile, it is useful briefly to introduce two more functional relationships as a

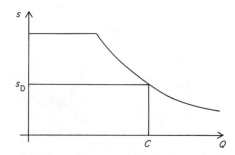

Figure 7.16 Speed-flow relationship

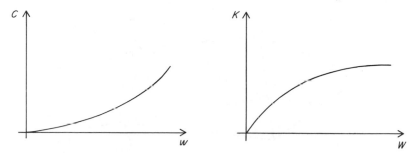

Figure 7.17 (a) Capacity–width relationship; (b) cost–width relationship

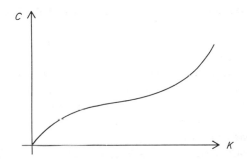

Figure 7.18 Capacity–cost relationship

prelude to another discussion below—on hierarchies. As we have seen, the obvious notion of the *capacity* of a link relates to the flow it can carry. It is useful to relate this concept of capacity to measures of physical size and cost of a link. Again, for the sake of illustration, we can stick to road links. Suppose we measure physical size by the width, W, of a link. Let C be the capacity as before, and let K be the capital cost (for a given length of link). Then we would expect these quantities to be related to W as shown in Figure 7.17(a) and (b). That is, capacity will increase more rapidly than linearly with width, while capital cost increases less than linearly: this is a graphical representation of scale economies. This implies a relationship between C and K as shown in Figure 7.18. These relationships provide the basis for a discussion of trunk concepts in the context of hierarchies in Section 7.2.3.

7.2.2 Shortest paths

To fix ideas, we begin with the network already presented as Figure 7.11 and concentrate on link travel times only. Thus, we are seeking shortest *time* paths in the network for the present. The complexities of generalized cost will be added later. The network with travel times shown is exhibited as Figure 7.19. Suppose we now focus on node 1: inspection shows that the shortest paths from that node

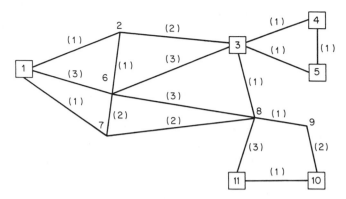

Figure 7.19 Example network, with link costs

to all other nodes are exhibited by Figure 7.20. This has a tree-like structure (no circuits) as predicted earlier. In more realistic networks, it is difficult or impossible to identify the shortest paths by inspection, and computer algorithms are needed for this purpose. One of the best known is that due to Dantzig, and this is described below. The details of this can be omitted without too much loss of understanding, however, for the reader who is prepared simply to accept that such computer algorithms exist.

The algorithm can be described as follows. Let nodes be denoted by indices such as r and s and let the link travel time be t_{rs}. Each node, s, say is given an additional pair of labels $[r, t(s)]$*. Let r_0 be the node from which the shortest paths are to be found. Then, for each $s \neq r_0$, the label is set to $[r_0, \infty]$ initially, and for $s = r_0$, the lebel is $[r_0, 0]$. The heart of the algorithm then consists of the following steps:

(i) Set $r = r_0$ initially.
(ii) Search for a link (r, s) such that

$$t(r) + t_{rs} < t(s). \tag{7.4}$$

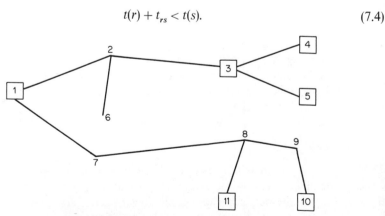

Figure 7.20 Shortest paths from node 1

*$t(s)$ is the minimum time for getting to s from r.

Table 7.1

	Starting values of labels	Choose 1	Choose 2
1	[1, 0]	[1, 0]	[1, 0]
2	[1, ∞]	[1, 1]*	[1, 1]
3	[1, ∞]	[1, ∞]	[2, 3]*
4	[1, ∞]	[1, ∞]	[1, ∞]
5	[1, ∞]	[1, ∞]	[1, ∞]
6	[1, ∞]	[1, 3]*	[2, 2]*
7	[1, ∞]	[1, 1]*	[1, 1]
8	[1, ∞]	[1, ∞]	[1, ∞]
9	[1, ∞]	[1, ∞]	[1, ∞]
10	[1, ∞]	[1, ∞]	[1, ∞]
11	[1, ∞]	[1, ∞]	[1, ∞]

	Choose 7	Choose 6	Choose 3
1	[1, 0]	[1, 0]	[1, 0]
2	[1, 1]	[1, 1]	[1, 1]
3	[2, 3]	[2, 3]	[2, 3]
4	[1, ∞]	[1, ∞]	[3, 4]*
5	[1, ∞]	[1, ∞]	[3, 4]*
6	[2, 2]	[2, 2]	[2, 2]
7	[1, 1]	[1, 1]	[1, 1]
8	[7, 3]*	[7, 3]	[7, 3]
9	[1, ∞]	[1, ∞]	[1, ∞]
10	[1, ∞]	[1, ∞]	[1, ∞]
11	[1, ∞]	[1, ∞]	[1, ∞]

	Choose 8	Choose 4	Choose 5
1	[1, 0]	[1, 0]	[1, 0]
2	[1, 1]	[1, 1]	[1, 1]
3	[2, 3]	[2, 3]	[2, 3]
4	[3, 4]	[3, 4]	[3, 4]
5	[3, 4]	[3, 4]	[3, 4]
6	[2, 2]	[2, 2]	[2, 2]
7	[1, 1]	[1, 1]	[1, 1]
8	[7, 3]	[7, 3]	[7, 3]
9	[8, 4]*	[8, 4]	[8, 4]
10	[1, ∞]	[1, ∞]	[1, ∞]
11	[8, 6]*	[8, 6]	[8, 6]

	Choose 9
1	[1, 0]
2	[1, 1]
3	[2, 3]
4	[3, 4]
5	[3, 4]
6	[2, 2]
7	[1, 1]
8	[7, 3]
9	[8, 4]
10	[9, 6]*
11	[8, 6]

No further improvement is obtained by choosing nodes 10 or 11.

* = change from previous round

(iii) If such a link is found, change the label on node s to $[r, t(r) + t_{rs}]$.
Continue steps (ii) and (iii) until no such links are found. Then
(iv) Take any node q, say, which has not played the role of r in steps (ii) and (iii) and for which $t(q)$ is a minimum.
Repeat steps (iii) and (iv) for this node q, and repeat this process until no such new nodes q can be found.

The only way to understand this rather abstract description of an algorithm is to follow it through for a particular example. This is done in Table 7.1 for the

Table 7.2

	Starting values of labels	Choose 8	Choose 3
1	$[8, \infty]$	$[8, \infty]$	$[8, \infty]$
2	$[8, \infty]$	$[8, \infty]$	$[3, 3]^*$
3	$[8, \infty]$	$[8, 1]^*$	$[8, 1]$
4	$[8, \infty]$	$[8, \infty]$	$[3, 2]^*$
5	$[8, \infty]$	$[8, \infty]$	$[3, 2]^*$
6	$[8, \infty]$	$[8, 3]^*$	$[8, 3]^*$
7	$[8, \infty]$	$[8, 2]^*$	$[8, 2]$
8	$[8, 0]$	$[8, 0]$	$[8, 0]$
9	$[8, \infty]$	$[8, 1]^*$	$[8, 1]$
10	$[8, \infty]$	$[8, \infty]$	$[8, 0]$
11	$[8, \infty]$	$[8, 3]^*$	$[8, 3]$

	Choose 9	Choose 4	Choose 5
1	$[8, \infty]$	$[8, \infty]$	$[8, \infty]$
2	$[3, 3]$	$[3, 3]$	$[3, 3]$
3	$[8, 1]$	$[8, 1]$	$[8, 1]$
4	$[3, 2]$	$[3, 2]$	$[3, 2]$
5	$[3, 2]$	$[3, 2]$	$[3, 2]$
6	$[8, 3]$	$[8, 3]$	$[8, 3]$
7	$[8, 2]$	$[8, 2]$	$[8, 2]$
8	$[8, 0]$	$[8, 0]$	$[8, 0]$
9	$[8, 1]$	$[8, 1]$	$[8, 1]$
10	$[9, 3]^*$	$[9, 3]$	$[9, 3]$
11	$[8, 3]$	$[8, 3]$	$[8, 3]$

	Choose 7	Choose 2	
1	$[7, 3]^*$	$[2, 4]^*$	
2	$[3, 3]$	$[3, 3]$	
3	$[8, 1]$	$[8, 1]$	
4	$[3, 2]$	$[3, 2]$	No further
5	$[3, 2]$	$[3, 2]$	improvement is
6	$[8, 3]$	$[8, 3]$	obtained by
7	$[8, 2]$	$[8, 2]$	choosing nodes
8	$[8, 0]$	$[8, 0]$	6, 10, 11, or 1.
9	$[8, 1]$	$[8, 1]$	
10	$[9, 3]$	$[9, 3]$	
11	$[8, 3]$	$[8, 3]$	

Figure 7.19 network, choosing node 1 as the origin zone. The reader should check the way in which each successive column in produced according to the algorithm. Once the final column of labels is produced, the shortest path from node 1 to any other node is obtained by tracing back the first part of the label. Consider node 11 for example, which has the final label [8, 6]. This shows that it can be reached via node 8 from node 1 (in 6 time units in all). The node 8 entry is [7, 3], which shows that it is reached via node 7; the node 7 entry is [1, 1], which connects to node 1 directly. Thus the shortest path from node 1 to node 11 is (1, 7, 8, 11), and the total time taken is 6 units. The final column, therefore, is a representation of the tree shown in Figure 7.20.

The workings of the algorithms for node 8 as origin zone are displayed in Table 7.2 and the tree structure implied by the final column is shown in Figure 7.21.

It is a straightforward matter, of course, to build shortest paths of link properties other than travel time. t_{rs} above can be replaced by another measure — or combinations of measures if generalized costs are involved. Any nodal costs can also be added in. Thus, if γ_{rs} are link costs, and $\gamma_i^{(O)}$ and $\gamma_j^{(D)}$ are origin and destination costs at nodes i and j respectively, then the generalized cost c_{ij}, from source i to sink j, which we have used so much in other chapters, can now be related to underlying network costs by the formula

$$c_{ij} = \gamma_i^{(O)} + \sum_{(r, s)\in R_{ij}^{\min}} \gamma_{rs} + \gamma_j^{(D)} \tag{7.5}$$

where $R_{ij}^{\min}$ is the set of links forming the shortest path between i and j and so $\sum_{(r, s)\in R_{ij}^{\min}} \gamma_{rs}$ is simply an appropriate notation for the sum of link cost on the shortest route between i and j. Thus, this network basis for inter-*zonal* costs (or times or distances) should always be borne, in mind. There is also a further complication which arises from the interdependence of link travel times and flows and this is developed in the context of a discussion of network assignment in the next subsection.

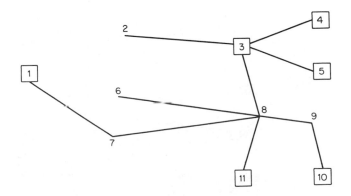

Figure 7.21 Shortest paths from node 8

Table 7.3

	1	3	4	5	10	11
1	X	102	32	X	29	53
3	X	X	350	X	203	151
4	X	X	X	X	X	X
5	X	X	X	X	X	X
10	X	X	X	X	X	X
11	X	X	X	X	X	X

7.2.3 Flow assignment

Assignment is concerned with loading origin–destination (i.e. source–sink) flows on to a network. Initially, we make the assumption that such flows are loaded on to shortest paths. Origin–destination flows take the form of an array $\{T_{ij}\}$, where i and j are possible sources and sinks and we have seen that link flows can be conveniently denoted by Q_{rs}, where r and s are nodes. A full description of the assignment of flows to links of the network would involve a four-dimensional array, say $\{x_{ij}^{rs}\}$—the contribution to load on link (r, s) from an (i, j) flow.

To fix ideas, consider the origin–destination table given in Table 7.3 for the sources and sinks of the network given in Figure 7.11 (or 7.19). Only seven flows have been identified for simplicity and given a key as shown.

The 102 units flowing from 1 to 3 have to be loaded on links $(1, 2)$ and $(2, 3)$ since they from the shortest parth from 1 to 3. The loading of all seven flows, in this way, is shown in Figure 7.22. It can easily be seen how this principle of assignment could be extended if more flows in the table were identified explicitly, or for bigger networks and tables.

There are 18 nonzero x_{ij}^{rs} elements identified on Figure 7.22. For example, $x_{13}^{12} = 102$ and $x_{14}^{12} = 32$ are the two contributions to the loads on link $(1, 2)$,

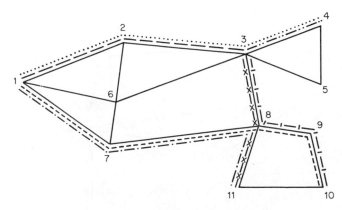

Figure 7.22 Flows assigned to shortest paths

making a load of 134 in all. In general, the load on a link is

$$Q_{rs} = \sum_{ij} x_{ij}^{rs} \qquad (7.6)$$

(though most of the x_{ij}^{rs} will be zero for any particular (r,s)). In the case of the example cited, the nonzero contributions are

$$Q_{12} = x_{13}^{12} + x_{14}^{12} = 102 + 32 = 134. \qquad (7.7)$$

The assignment (or network loading) process described here can easily be programmed for a computer and very large networks and origin–destination tables can be handled in this way. However, the method described above only works if certain rather restrictive conditions hold—in particular, if the link times are independent of the flows and the link loadings. The more difficult problem when this is not true is considered in the next subsection.

7.2.4 Flow-link time interdependence: network equilibrium

We continue with the transport example to illustrate the effects of congestion in networks, though the idea is applicable to a wider range of problems as we will see later. First, recall that the interchange flows, from sources to sinks, $\{T_{ij}\}$, are a function of costs, c_{ij}:

$$T_{ij} = T_{ij}(c_{ij}, \text{ other variables}) \qquad (7.8)$$

(using an obvious notation). Secondly, c_{ij} is a sum of link costs:

$$c_{ij} = \sum_{(r,s) \in R_{ij}^{min}} \gamma_{rs} \qquad (7.9)$$

say (neglecting origin or destination costs for now) But we also saw that link costs (at least through their time component) are functions of link flows, say:

$$\gamma_{rs} = \gamma_{rs}(Q_{rs}, \text{ other variables}). \qquad (7.10)$$

Finally, we recall that the analysis of the previous subsection implies that

$$Q_{rs} = \sum_{(i,j) \in V_{rs}^{min}} T_{ij} \qquad (7.11)$$

where V_{rs}^{min} is the set of (i,j) pairs for which (r,s) is on the shortest path. (This is a way of writing equation (7.6) without having to define x_{ij}^{rs} explicitly.)

Equations (7.8)–(7.11) now exhibit an analytically difficult circularity: T_{ij}'s are functions of c_{ij}'s; c_{ij}'s are functions of γ_{rs}'s; γ_{rs}'s are functions of Q_{rs}'s; and Q_{rs}'s are functions of T_{ij}'s. The restricting assumptions of the previous subsection broke the circularity by assuming that $\{\gamma_{rs}\}$ was given and fixed—say at free-flow values—and not dependent on $\{Q_{rs}\}$ and hence on $\{T_{ij}\}$. In general, of course, such restrictions are unrealistic: they will only hold for networks carrying very little traffic (and if the American time-flow relationship (7.2) holds, not even then).

When link flows and times are interdependent, the network is said to be congested.

As we have seen in other contexts, the modellers' escape from this kind of strong circular interdependence is to use iteration. An obvious scheme here is:

(i)   Assume an initial set of γ_{rs}'s (either free-flow or 'observed') and hence c_{ij}'s.
(ii) Calculate $\{T_{ij}\}$ from (7.8).
(iii) Calculate $\{Q_{rs}\}$ from (7.11).
(iv) Calculate $\{\gamma_{rs}\}$ from (7.10).
(v)   Calculate new $\{c_{ij}\}$ from (7.9).
(vi) Repeat from step (ii) until convergence is achieved.

In practice, such convergence usually is achievable and the result is an *equilibrium* state of the congested network.

If this procedure was applied to the network of Figure 7.19 and the associated origin–destination Table 7.3, then as an additional preliminary, a time-flow relation would have to be associated with each link of the network — say equation (7.2) with constants chosen as appropriate for each link. Suppose then that Table 7.3 was the outcome of a $\{T_{ij}\}$ model for the initial c_{ij}'s and γ_{rs}'s. We would proceed through step (iii) above and find $\{Q_{rs}\}$ by loading $\{T_{ij}\}$ onto the network (as described in the previous subsection). At step (iv) it would then be discovered that when Q_{rs}, for each (r, s), was substituted into the time-flow equation, the γ_{rs} was different to the one used in the first place. The iteration continues until this is no longer true (within some given tolerance) and it is then said that convergence has been achieved. Note that implicit in the scheme described in steps (i)–(vi) above is that, at step (v) when new $\{c_{ij}\}$ are obtained, new minimum paths have to be found — a new application of Dantzig's algorithm is needed here at each cycle of the iteration. This is an important point to bear in mind for the next part of the discussion.

The method described above is known as 'all-or-nothing' congested assignment because at the end of the iterative cycle, each (i, j) flow is assigned to only one path — 'all or nothing'. This mechanism is unlikely to reflect real life: if near-substitutes are available, some people will probably use them. Various methods have been used to produce multi-path assignment techniques. A detailed description would take us beyond the scope of this book but some brief sketches of ideas may be useful. The basic problem is that, while an algorithm can be stated for producing shortest (or best) paths in networks, no similar algorithm is known for finding second, third, or nth, best paths. Thus, some approximate scheme has to be devised. For example, the iterative scheme described above does, as we noted, produce (possibly) different best paths at each stage in the iteration. For each (i, j) pair, the set of paths produced during the iteration are likely to be good substitutes for each other, and these are sometimes then used as the basis for multi-path procedures. One such procedure is a neat variant on this idea: so-called incremental loading. In this method, some fraction of the origin–destination table is loaded onto the networks, then link times are adjusted, then another fraction loaded, and so on until all the trips are loaded. This achieves multi-path assignment but still has to be embedded in an overall iterative procedure which involves the recalculation of $\{T_{ij}\}$ based on the link times obtained at the end of the incremental procedure. The more formal basis of these ideas, which ties up with aspects of entropy maximizing methods (Chapter 4) and mathematical programming methods (Chapter 6), is sketched in the next subsection.

7.2.5 Mathematical programming and network equilibrium

We saw in Chapter 4 that the distribution model equations for $\{T_{ij}\}$ could be derived from entropy maximizing methods, or alternatively from group surplus maximization (based on random utility theory) in Chapter 6. If we choose the first method, we could state the problem very briefly as

$$\underset{\{T_{ij}\}}{\text{Max}} \, S = -\sum \log T_{ij}! \tag{7.12}$$

subject to

$$\text{'entropy constraints'} = 0 \tag{7.13}$$

where we use an extreme abbreviation for the constraints. Suzanne Evans (1976) has shown that the probelm

$$\underset{\{Q_{rs}, T_{ij}\}}{\text{Max}} \, Z = -\sum_{rs} \Gamma_{rs}(Q_{rs}) - \frac{1}{\beta} \sum_{ij} \log T_{ij}! \tag{7.14}$$

(for a suitable function Γ_{rs}) subject to

$$\text{'entropy constraints'} = 0 \tag{7.15}$$

and

$$\text{'network constraints'} = 0 \tag{7.16}$$

produces an entropy maximizing $\{T_{ij}\}$ and flows, $\{Q_s\}$, which satisfy 'all-or-nothing' congested equilibrium conditions. The constraints have been given in this abbreviated form because more detail would take us substantially beyond the scope of this book, but essentially they represent the 'connectedness' of routes and balance equations (Kirchhoff's Laws) at nodes.

We can, however, sketch the nature of the function Γ_{rs} as this leads to new insights and also allows us to introduce some older concepts associated with assignment—Wardrop's principles. The *first principle* of Wardrop (1952) described a system within which individuals each minimized their individual travel times—and all the assignment algorithms given in the preceding two subsections satisfy such a rule. This is appropriate for the traffic problem because because that is how we would except people to behave. An alternative principle, Wardrop's second, is that traffic should be assigned so that total systemwide cost is minimized. This may be feasible for a network in which all the flows were controlled by a single authority—say in the assignment of resource flows (petrol tankers?) to a network. We now find that investigation of Evans' Γ_{rs} function leads to an interesting result (an extension by her of earlier work by Potts and Oliver).

$\Gamma_{rs}(Q_{rs})$ is defined by

$$\Gamma_{rs}(Q_{rs}) = \int_0^{Q_{rs}} \gamma_{rs}(q)\,dq \qquad (7.17)$$

where $\gamma_{rs}(q)$ is the travel time on link (r, s) when the link flow is Q. For those who do not understand the definition using the integral, and to produce a result which is useful anyway, we note that Γ_{rs} is related to an average cost, $\bar{\gamma}_{rs}$, defined by

$$\bar{\gamma}_{rs} = \frac{1}{Q_{rs}} \int_0^{Q_{rs}} \gamma_{rs}(q)\,dq. \qquad (7.18)$$

The first term in the summation (7.14) can then be written

$$-\sum_{rs} Q_{rs}\bar{\gamma}_{rs} \qquad (7.19)$$

and this shows that total systemwide costs *are* being minimized provided these costs are taken as $\bar{\gamma}_{rs}$. Thus, while Wardrop's first principle is satisfied in relation to the 'real' costs, $\gamma_{rs}(q)$, rather oddly, his second principle is satisfied in relation to average costs, $\bar{\gamma}_{rs}$.

The final step in this brief mathematical exploration of the main results is to connect back to another part of Chapter 6 where the basic distribution models were disaggregated. In the transport case, this meant disaggregation by mode and person type. An important step was the introduction of equation (6.101) (or (6.102)) to relate model costs, c_{ij}^k, to the more aggregate level, c_{ij}^n. Now that we have seen the possibility of multi-route assignment, we can note the task of relating c_{ij}^k to the more disaggregate route costs, say a_{ij}^{kl} for the lth route, mode k, between i and j. Exponential averaging is

$$e^{-\mu^k c_{ij}^k} = \sum_l e^{-\mu^k a_{ij}^{kl}} \qquad (7.20)$$

where μ^k is the *route split parameter* in

$$R_{ij}^{kl} = \frac{e^{-\mu^k a_{ij}^{kl}}}{\sum_l e^{-\mu^k a_{ij}^{kl}}}. \qquad (7.21)$$

Unfortunately, within the random utility framework, the relationships are more complicated. This arises because routes often have links in common and this leads to the problem of handling attribute correlation. The reader is referred to the work of Williams in 'Notes on further reading' below for further investigation on this.

7.2.6 Connectedness 1: elementary approaches

We have already noted (in relation to Figure 7.6) that it can be seen by observation that some networks are wholly 'connected', and some not. We begin

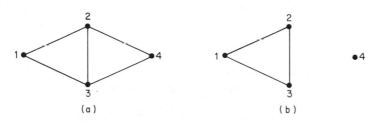

(a)   (b)

Figure 7.23 Complete and incomplete networks

this section by formalizing this idea using elementary matrix algebra and then indicate deeper approaches to the analysis of connectedness and related aspects of network structure. The first part of the presentation relies heavily on the book by Roberts (1976) and the account of q-analysis on the work of Atkin (1974, 1977), both of which are strongly recommended for further reading.

Consider a network such as that shown in Figure 7.23(a). The structure of the network can be represented in an *adjacency matrix*: let A be the set of links, in this case $\{(1, 2), (2, 3), (1, 3), (2, 4), \text{and} (3, 4)\}$, and let a_{ij} be 1 if (i, j) is a link and be zero if not. The matrix **a** is then

$$\mathbf{a} = \{a_{ij}\} = \begin{bmatrix} 0 & 1 & 1 & 0 \\ 1 & 0 & 1 & 1 \\ 1 & 1 & 0 & 1 \\ 0 & 1 & 1 & 0 \end{bmatrix}.$$

(7.22)

It has been assumed that the links are two-way, and so the matrix is symmetric. a_{ii} is set to zero for each i. Consider the matrices $\mathbf{a}^2, \mathbf{a}^3$, and $\mathbf{a}^4$ (recalling that $(\mathbf{ab})_{ij} = \sum_k a_{ik} b_{kj}$):

$$\mathbf{a}^2 = \begin{bmatrix} 2 & 1 & 1 & 2 \\ 1 & 3 & 2 & 1 \\ 1 & 2 & 3 & 1 \\ 2 & 1 & 1 & 2 \end{bmatrix}$$

(7.23)

$$\mathbf{a}^3 = \begin{bmatrix} 2 & 5 & 5 & 2 \\ 5 & 4 & 5 & 5 \\ 5 & 5 & 4 & 5 \\ 2 & 5 & 5 & 2 \end{bmatrix}$$

(7.24)

$$\mathbf{a}^4 = \begin{bmatrix} 10 & 9 & 9 & 10 \\ 9 & 15 & 13 & 9 \\ 9 & 13 & 15 & 9 \\ 10 & 9 & 9 & 10 \end{bmatrix}.$$

(7.25)

The a_{ij}- element of $\mathbf{a}^n$ contains the number of paths between i and j of length n. In this rather trivial example, the alternative paths (of which there seem to be quite high numbers in the higher-powered matrices) arise from traversing links backwards and forwards and the large number of combinations thus generated.

Now consider the matrix

$$\mathbf{b} = \mathbf{I} + \mathbf{a} + \mathbf{a}^2 + \mathbf{a}^3 + \mathbf{a}^4 \tag{7.26}$$

(where $\mathbf{I}$ is the unit matrix and is added to indicate that a node is considered reachable from itself). The matrix $\mathbf{R}$ is constructed from $\mathbf{b}$ by the rule

$$R_{ij} = \begin{cases} 1 & \text{if } b_{ij} > 0 \\ 0 & \text{if } b_{ij} = 0. \end{cases} \tag{7.27}$$

Then, for the example

$$\mathbf{R} = \begin{bmatrix} 1 & 1 & 1 & 1 \\ 1 & 1 & 1 & 1 \\ 1 & 1 & 1 & 1 \\ 1 & 1 & 1 & 1 \end{bmatrix} \tag{7.28}$$

$\mathbf{R}$ is known as the reachability matrix: $R_{ij} = 1$ indicates that j is reachable from i; $R_{ij} = 0$ that this is not the case. In our example, the matrix is shown to be wholly connected. If the analysis is repeated for the network in Figure 7.23(b), we would get

$$\mathbf{R} = \begin{bmatrix} 1 & 1 & 1 & 0 \\ 1 & 1 & 1 & 0 \\ 1 & 1 & 1 & 0 \\ 0 & 0 & 0 & 1 \end{bmatrix} \tag{7.29}$$

showing that the node 4 is reachable only from itself and is disconnected from the rest of the network. In more complicated networks, especially where many of the internodal links are one-way only, results on connectedness are not obvious to the naked eye and this kind of matrix analysis is invaluable.

This kind of formal analysis of connectedness can be taken further — for example by investigating the relative contribution of different links and developing a measure of *vulnerability* which shows how much the degree of connectedness of a network is worsened when a particular link is removed. However, this amount of detail takes us beyond the scope of this book and the reader is referred to the book by Roberts as listed in the 'Notes on further reading' section below.

The next two steps in the argument highlight the application of different kinds of network theory to more abstract structures of systems — though some of the techniques applied mainly to physical networks above are applicable in this context, and vice versa. The first is a continuation of the earlier argument and is concerned with directed graphs, or *digraphs*. The second involves a different kind of mathematics and is known as *q*-analysis.

7.2.7 Connectedness 2: digraphs

When a digraph is used as a representation of a system, the nodes in the graphs are 'stores' (c.f. Chapter 2), each containing a number of system components of

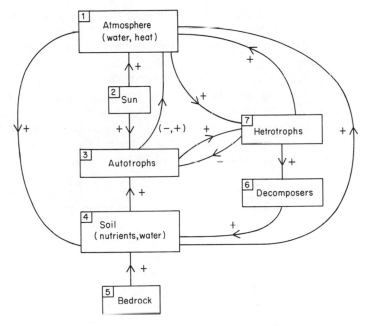

Figure 7.24 A signed digraph representation of an ecosystem

some type. The links represent causal effects, so that (i, j) is a representation of the effect of a change in i on j, and this, of course, implies a direction. Consider, for example, the ecosystem diagram, first introduced as Figure 2.11 in Chapter 2, which can be taken as a representation of our moorland ecosystem at a coarse level of resolution. Let the main units be considered as stores, and we can add links to show directions of change as in Figure 7.24. This is shown in a more

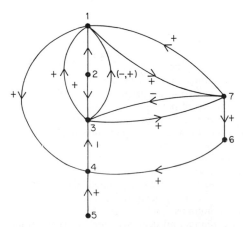

Figure 7.25 More abstract representation of an ecosystem

abstract, and conventional, digraph from in Figure 7.25. The flows represented by the links are listed in Table 7.4.

The signs are all clearly positive except two: the link (7, 3) represents the effect of herbivores eating the vegetation; and a link (3, 2) has been included to indicate that, in the early stages of growth, plants will be able to process more energy, and so can 'receive' more — though this is a link to represent a mechanism rather than a flow in a direct sense; in later stages, the canopy will block incident radiation from lower-standing vegetation, and the top, in the case of heather, cannot process radiation sufficiently rapidly to achieve further growth, and so this is a representation of a negative effect — that which creates the *cycle* of growth of heather.

The main cycles can be easily identified: water, nutrients, and energy. The system is driven by the energy supply at store 2 and the nutrient supply at store 5. It is not surprising, therefore, that most of the cycles represent positive feedback loops. The only negative loops are (2, 3, 2) — in the later stages of growth — and (3, 7, 3), which represents the prey–predator relation between animals and vegetation (which will be discussed from another perspective in Chapter 8, building on its introduction to illustrate state space in Chapter 2). In this case, therefore, as we know from experience of moorland ecosystems, most system components will grow for quite long periods (with the life-cycle of heather being about 30 years) but will ultimately be checked by the impact of negative feedback loops — either the heather cycle itself or the effects of grazing.

These kinds of results are typical of digraph analysis and are available in a general form. For example, the nature of feedback loops can be identified as follows: if the number of minus signs in a loop is odd, the feedback is negative; otherwise, positive. It also seems to be the case that the greater the number of 'stores' (i.e. types of components) there are in the system, the greater is the chance that the overall behaviour of the system will be stable.

Table 7.4

	Flow
(1, 4)	Water from atmosphere to soil
(1, 7)	Water from atmosphere to animals
(2, 1)	Energy from sun to atmosphere
(2, 3)	Energy from sun of plants
(3, 1)	Energy from plants to atmosphere
(3, 2)	Increasing canopy with plant growth reduces incident radiation and energy processing capabilities
(3, 7)	Plants to herbivores
(4, 3)	Nutrients and water to plants
(5, 4)	Nutrients from bedrock weathering to soil
(6, 4)	Nutrients produced by decomposers to soil
(7, 1)	Energy from animals
(7, 3)	Animals eat plants and decrease stock of vegetation
(7, 6)	Dead animals to be decomposed

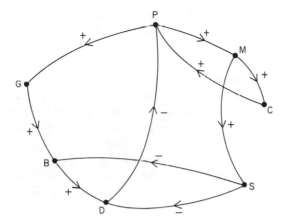

Figure 7.26 Solid waste disposal in an area: a signed digraph of mechanisms (after Roberts, 1976, p187, Figure 4.6).

G: Amount of garbage area   S: Sanitation facilities

P: Population in a city   D: Number of diseases

M: Modernization   B: Bacteria/area

C: Migration into a city

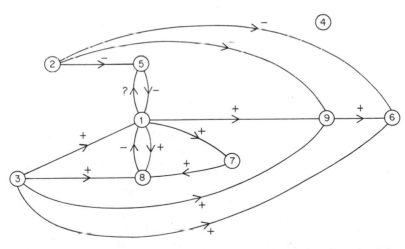

Figure 7.27 A signed digraph of passenger transport relations in a city (after Roberts, 1976, p189, Figure 4.8).

(1) Passenger miles   (5) Price of commuter ticket
(2) Fuel economy   (6) Emissions
(3) Population size   (7) Accidents
(4) Cost of car   (8) Possibility of delay
    (9) Fuel consumption.

158

Most systems would have more minus signs in their digraphs than was the case in the ecosystem example above. To fix ideas, two more examples, taken from Roberts' book, are presented in Figure 7.26 and 7.27, both concerned with aspects of the city. The first is based on the work of Maruyama (1963) and is concerned with facilities for solid waste disposal; the second is from Roberts' own work and describes the volume, and related effects, of intra-urban car commuting. The reader will be able to develop many more examples of his own.

The next step in this argument is to quantify in some way the changes implied by links and this leads to the concept of a *weighted digraph*. An example from Roberts is shown in Figure 7.28. The weights give the change in the 'receiving' store for a unit change in the originating one in some time interval, δt. For example, if $u_2(t)$ and $u_9(t)$ are the store totals at 2 and 9 at time t, the figure shows that

$$u_9(t + \delta t) = u_9(t) - 0.9[u_2(t + \delta t) - u_2(t)]. \tag{7.30}$$

More generally, the weights could be *functions* of the various totals and we could sum various effects together:

$$u_j(t + \delta t) = u_j(t) + \sum_i w_{ij}(u_i(t), u_j(t)) \times [u_i(t + \delta t) - u_i(t)] \tag{7.31}$$

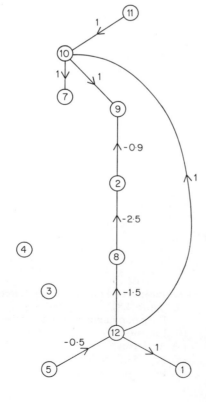

Figure 7.28 Energy use and clean air in an urban transportation system: a signed digraph (after Roberts, 1976, p239, Figure 4.50).
(1) Passenger miles
(2) Fuel economy of cars in m.p.g.
(3) Population size
(4) Cost of car
(5) Price of trip
(6) Emissions
(7) Accidents
(8) Average delay
(9) Fuel consumption
(10) Vehicle miles
(11) Number of cars
(12) Number of person trips.

where the w_{ij}'s are the weights (and of course are zero if there is no link between a particular i and j). It will easily be seen from the presentation of the next chapter that sets of equations of this type are sets of difference equations and that the method is closely related to the 'systems dynamics' approach of Forrester.

7.2.8 Connectedness 3: q-analysis

The method known as q-analysis has been developed by Atkin, and this section is based entirely on the work of him and his collaborators. It is based on the notion of *sets* and *relations* between elements of sets and perhaps does not sit entirely comfortably in a chapter on network analysis, but the topological structures generated by it look in some respects like networks and seem to generate insights into system behaviour which fall into the same family.

It is best to proceed by example. Let $X1, X2, X3, \ldots$ be the components of a system and form the set X; let $A1, A2, \ldots$ be the activities of these components. In the interesting cases, there is *not* a $1-1$ correspondence between components (or component types) and activities and the relation connecting can be shown in a matrix (which is rather like the adjacency matrix of the previous section). To fix ideas, consider the system defined by Figure 7.25. For convenience, the components are listed again in Table 7.5 below together with an appropriate set of 'activities'.

Let the relation λ be defined between sets X and A as in the matrix, say Λ, in Table 7.6 below. The assumed relationships between system components and activities are mostly self explanatory. Perhaps the only real oddity is that the sun, $X2$, is shown as an energy consumer. This is a formal device to represent the effect of plant growth on reducing effective per unit biomass energy utilization. Both autotrophs and heterotrophs are shown as animal producers (to allow for herbivores and carnivores).

The investigation of system structure on this basis proceeds as follows. Each

Table 7.5

	Components		Activities
$X1$.	Atmosphere	$A1$.	Energy production
$X2$.	Sun	$A2$.	Energy consumption
$X3$.	Autotrophs	$A3$.	Water production
$X4$.	Soil	$A4$.	Water consumption
$X5$.	Bedrock	$A5$.	Vegetation production
$X6$.	Decomposers	$A6$.	Vegetation consumption
$X7$.	Heterotrophs	$A7$.	Animal production
		$A8$.	Animal consumption
		$A9$.	Heat production
		$A10$.	Heat consumption
		$A11$.	Nutrient production
		$A12$.	Nutrient consumption

Table 7.6

	A1	A2	A3	A4	A5	A6	A7	A8	A9	A10	A11	A12	q
X1			1	1						1			2
X2	1	1											1
X3			1		1		1		1			1	4
X4				1	1							1	2
X5				1							1		1
X6							1	1		1			2
X7				1		1	1	1	1				4
	0	1	0	4	0	0	1	1	2	0	1	1	

system component is represented by a set of points in activity space (which, as a state-space, is twelve dimensional). Thus $X1$ is represented by three points $(A3, A4, A10)$; since three points form a plane, this is a two-dimensional object and is known as a 2-simplex. It is shown diagrammatically in Figure 7.29. $X2$ is simply a line in this representation and so is a 1-simplex. In general, a system component is represented as a q-simplex; hence the name of the method. The q-values of each simplex are shown in the final column of Table 7.6.

The whole set of simplexes is known as a simplicial complex. Since this is a list of X objects represented in A-space, and based on the relation λ, it is denoted by $K_X(\lambda, A)$. It is written out in full in Table 7.7.

A q-analysis consists of taking a simplicial complex, such as that in Table 7.7, and identifying, in decreasing order, the objects in activity space. Thus, $X3$ will be at the top of the list as a 4-simplex. But the analysis also includes new objects formed when two or more simplexes have points in common. The q-analysis for Table 7.7 is shown in Table 7.8. The various objects thus identified are shown in Figure 7.30. They can be slotted together in a geometrical structure as shown in Figure 7.31.

It can be argued that the picture produced in Figure 7.31 is revealing in a number of ways. It shows the close interdependence of $X3$ and $X7$—the autotrophs and heterotrophs—and the relative independence of the various 'driving' components and associated activities—$X1$ (the atmosphere), $X2$ (the sun), $X5$ (the bedrock), and, to a lesser extent, of $X4$ (the soil), and $X6$ (the

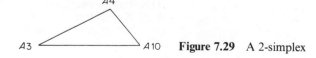

Figure 7.29 A 2-simplex

Table 7.7

$X1$	$\{A3, A4, A10\}$
$X2$	$\{A1\}$
$X3$	$\{A2, A4, A7, A9, A12\}$
$X4$	$\{A4, A5, A12\}$
$X5$	$\{A4, A11\}$
$X6$	$\{A8, A9, A11\}$
$X7$	$\{A4, A6, A7, A8, A9\}$

Table 7.8

q	
4	$X3, X7$
2	$X1, X4, X6, \{X3, X7\}$
1	$X2, X5, \{X3, X4\}, \{X6, X7\}$
0	$\{X5, X6\}, \{X1, X3, X4, X7\}$

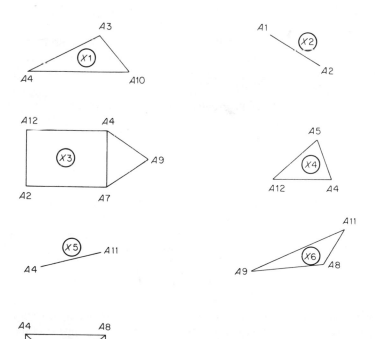

Figure 7.30 Elements of a q-analysis

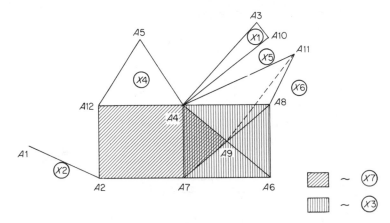

Figure 7.31   Connected elements of a q-analysis

decomposers). The various flows (heat/energy, water, nutrients) can be shown on these structures, as displayed in Figure 7.32. It is argued that the revealed structure has to 'carry' these flows and that q-analysis can either reveal 'holes' which impede flows, or can be used to assess the impact of dysfunction of an activity, say.

It is possible to carry out a conjugate analysis, using the columns of Table 7.6 instead of the rows. Each simplex is then an activity listed in the space of system components. The simplexes are listed in Table 7.9 and the resulting simplicial complex is labelled $K_A(\lambda^{-1}, X)$.

A q-analysis in this case produces the objects listed in Table 7.10. The corresponding geometrical representation is shown in Figure 7.33.

The analysis so far has been based on binary relations—the elements in the matrices such as Λ in Table 7.6 could only be zero or one. In many cases, a quantity can be associated with the relation—say the *volumes* of each activity associated with each component. q-analysis can still be carried out with this richer information base. The technique involves seeking new binary matrices using slicing parameters, usually with the possibility of a different parameter, say θ_i for each row—that is, each simplex. Thus, for a matrix $\{\mu_{ij}\}$ say, a binary matrix $\Lambda = \{\lambda_{ij}\}$ based on the slicing parameters $\theta = \{\theta_i\}$ defined by

$$\lambda_{ij} = \begin{cases} 1 \text{ if } \mu_{ij} > \theta_i \\ 0 \text{ otherwise.} \end{cases} \tag{7.32}$$

Obviously, a variety of such binary matrices can be constructed for various sets of slicing parameters. We will return to this discussion in the context of hierarchies below, where it has an obvious relevance.

The final step in this elementary introduction to q-analysis is the notion of a pattern. Each simplex in a complex may have a variety of properties or characteristics associated with it. Thus, if $X1, X2, \ldots$ are the simplexes, $\Pi(X1), \Pi(X2)$ may be properties, and Π is known as a *pattern* on $X1, X2, \ldots$. It is

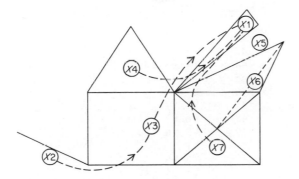

(a) Energy flows

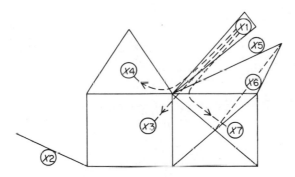

(b) Water flows

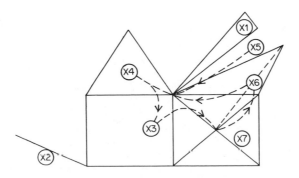

(c) Nutrient flows

Figure 7.32 A q-analysis of an ecosystem: (a) energy
flows; (b) water flows; (c) nutrient flows

Table 7.9

$A1$	$X2$
$A2$	$X2, X3$
$A3$	$X1$
$A4$	$X1, X3, X4, X5, X7$
$A5$	$X4$
$A6$	$X7$
$A7$	$X3, X7$
$A8$	$X6, X7$
$A9$	$X3, X6, X7$
$A10$	$X1$
$A11$	$X5, X6$
$A12$	$X3, X4$

Table 7.10

q	
4	$A4$
2	$A9$
1	$A2, A7, A8, A11, A12, (A4, A9)$
0	$A1, A3, A5, A6, A10$
	$\{A3, A4, A10\}, \{A1, A2\},$
	$\{A2, A4, A7, A9, A12\},$
	$\{A4, A5, A12\}, \{A4, A11\}$
	$\{A8, A9, A11\}, \{A4, A6, A7, A8, A9\}$

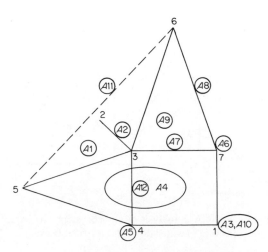

Figure 7.33 The dual q-analysis for the ecosystem

useful to arrange these in order of q-value in the complex. Let $\{X3, X4\}_n$, $\{X1, X6\}_{n-1}, \{X2, X8\}_{n-2}, \ldots$ for example be n-, $(n-1)$- and $(n-2)$-...simplexes. The pattern is then ordered in this way as $\{\Pi(X3), \Pi(X4)\}, \{\Pi(X1)\Pi(X6)\}$, $\{\Pi(X2), \Pi(X8)\}, \ldots$ which can be written in brief as $\Pi^n, \Pi^{n-1}, \Pi^{n-2}, \ldots$.

7.2.9 Hierarchies, nodes, and networks

The concept of 'hierarchy' will always figure prominently in any book on systems analysis. In this subsection, we are concerned with a relatively narrow aspect of the topic: the recognition of hierarchies in nodal and network structures. Three kinds of hierarchies, and techniques for identifying them, are reported briefly below. First, it is intuitively obvious that some nodes are more important than others and we find that this can be formalized by an examination of internodal flows. Secondly, some network links are more important than others, and we relate this notion to network order and to 'trunking'. Thirdly, we return to the more abstract higher-dimensional network structures generated by q-analysis and investigate hierarchies in that context.

To explore nodal hierarchies, we assume that a matrix of flows, say $\{T_{ij}\}$, between the elements of some set of nodes is available. Nystuen and Dacey (1961), in a well-known piece of work, showed how a nodal structure could be constructed from this. If D_j are the column sums of $\{T_{ij}\}$, that is the in-flows to $j, D_j = \sum_i T_{ij}$, then these are taken as a rank ordering of the relative importance of nodes. Then Nystuen and Dacey define a relation, say μ_{ij}, which is one if i and j are to be connected and zero otherwise. The rule for defining μ_{ij} (Wilson and Kirkby, 1975, p. 120) can be written as follows:

$$\mu_{ij} = 1 \tag{7.33}$$

if and only if

$$j = J_i \tag{7.34}$$

and

$$D_i > D_j \tag{7.35}$$

and

$$i \notin T \tag{7.36}$$

where J_i is the node which receives the maximum flow from i, and T is the set of terminals. i is a terminal if the largest flow from it is to a lower-order centre.

This rather abstract presentation of an algorithm can be described more informally as follows. First, identify the terminals, and then take each non-terminal, i, in turn, and connect it to *its* J_i. This will lead to each node being connected, directly or indirectly, to one terminal. The method produces a well-defined hierarchical structure: the terminals are at the top of the hierarchy, those nodes connected to terminals in the second rank, those connected to second-order nodes in the third, and so on. It should be emphasized that, as is often the case with concepts of this type, the rules given by (7.33)–(7.36) are relatively arbitrary and can be modified in various ways. Appropriate references are given in the 'Notes on further reading' section at the end of the chapter.

The 'links' generated by the above procedure are notional or abstract links which display hierarchical structure of *nodes* in networks. The next step is to look at network *links* directly. The simplest classification of the order of network links in tree-like networks, and especially river systems, is that due to Strahler (1965), building on earlier work by Horton and others. An example is shown in Figure 7.34. The principles of the scheme are evident from the figure: the lowest order links (or streams if we think of a river network for definiteness) are numbered 1; if two *n*-order streams meet, the 'product' is of order $n + 1$; if an *n*-order branch joins an $(n + 1)$-order stream, this remains (n + 1)-order. Thus, in the network shown in Figure 7.34, four orders are distinguished in a hierarchiccal scheme.

For the sake of clarity of presentation in the context of this book, other classification schemes are not presented here, though the interested reader can pursue them through the appropriate references which are signposted in the 'Notes on further reading' at the end of the chapter. It is more appropriate to raise in an exploratory fashion the alternative kinds of hierarchical network structures we may expect to find and to see why these may arise. Horton found that many networks could be described by a constant branching ratio, p. If N_i is the number of branches at level i, then

$$\frac{N_{i-1}}{N_i} = p \tag{7.37}$$

which turns out to imply that

$$N_i = p^{(k-2)} \tag{7.38}$$

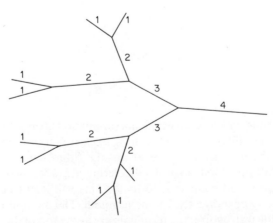

Figure 7.34 Strahler's classification of link orders in a river system

where k is the highest order in the basin ($k = 4$ in Figure 7.34, for example). Clearly, changing the value of p produces a variety of different network forms.

If we draw on the work of several authors, but particularly Woldenburg (1970), we can begin to sketch how hierarchical networks of this type come about. First, we observe that (in the river context, with appropriate analogues to be sought for other systems) a branch of a network serves an area—a branch of one order and its components of lesser orders serve a drainage basin. The whole system is made up of a nested set of such basins. Within any basin, the job of the river system is the transport of water which falls as rain evenly over the area. The 'least work' way of carrying the water is in channels, and the wider the channel, the less the friction cost per unit volume carried. However, the 'thinly spread' rain has to get to the channels. This can only be achieved efficiently with lower-order channels feeding higher-order channels, and hence the generation of a hierarchy. The nature of a particular hierarchy can then be seen as arising from (i) the size and topography of the area, (ii) the volume of water to be carried, and (iii) the relative efficiencies of different channel sizes. With the river system, the hierarchy arises from striking a balance (or equilibrium) between minimizing one set of costs (c_1) by carrying the greatest possible volume of water in the widest possible channel and minimizing another set of costs (c_2) involved in getting the water to the main channel. c_1 and c_2 are plotted against the width of the highest-order channel (W) in Figure 7.35. The sum curve $c = c_1 + c_2$ will have a clearly defined minimum as shown in Figure 7.36. A procedure of this kind would determine W_{opt}, therefore, and the numbers of lower-order branches (and the value of p or some such parameter) would arise from the other considerations mentioned of order and volume. Indeed, it may be that the analysis shown in Figure 7.35 and 7.36 could be carried out for successively smaller nested areas and thus determine the nature of the whole hierarchical system.

These considerations of network hierarchy do not apply only to tree-like networks. An obviously similar hierarchy arises in road networks, for example. One general concept underpinning both kinds of hierarchy is that of a *trunk channel*—a higher-order channel which has very much lower costs per unit than

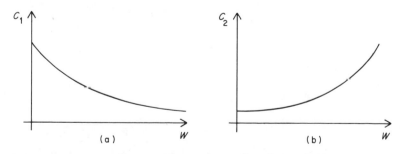

Figure 7.35 (a) c_1: 'cost' of transporting a unit of water versus channel width; (b) 'cost' of transporting a unit of water to next 'main' channel, versus width of subsiding channel

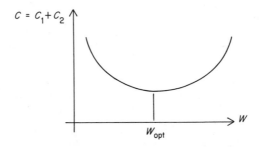

Figure 7.36 'Cost' minimization for channel width

lower-order channels. With river channels, these costs (or equivalently, capacities) may be expected to vary smoothly with changing channel width. With road (or some other) networks, there may be discontinuities. For example, when a road becomes 'big enough' to be a motorway, it is grade-separated from lower-order roads and this will lead to a discrete jump in its capacity relative to lower-order roads.

This argument can be summarized in another way. 'Trunking' comes about because the transport capacity (T) and the costs of providing it (C) increase with width as shown in Figure 7.35(a). This shows the existence of scale economies. Thus, 'trunk' links will be worthwhile if volumes are sufficiently large and *provided* the 'costs' of getting the water (or traffic, or whatever) to the trunk links are not too great.

The final step in our brief examination of hierarchies in the context of nodes and networks is to return to q-analysis. Atkin offers at least two approaches to hierarchy, the first one not yet discussed at all here; the second drawing on the presentation of the preceding subsection. The first arises from carefully distinguishing what we would call levels of resolution, and systems, subsystems, subsubsystems, etc. (with the possibility of components of one system being treated, as implied, as subsystems). Atkin's definition is rather broader, however: he builds on the notion of *cover sets* at different hierarchical levels, as exhibited for $N-1, N$, and $N+1$ in Figure 7.37. Atkin (1977) writes: 'If X is an N-level set and

$$N+1 \quad A \qquad\qquad B \quad \ldots$$

$$N \quad\quad X \quad Y \quad Z \quad \ldots$$

$$N-1 \quad PQR \ldots \ldots$$

$$A = \left\{ A_1 , A_2 , \cdots \right\}$$

$$A_1 = \left\{ X_1 , X_2 \right\}, \quad A_2 = \left\{ X_2 , X_3 , X_5 \right\}, \text{etc.}$$

$$X_1 = \left\{ P_2 , P_3 \right\}, \quad X_2 = \left\{ P_3 , P_6 \right\}, \text{etc.}$$

Figure 7.37 Hierarchical 'cover' in q-analysis

A a corresponding $(N + 1)$-level set then A must be a cover set for X, that is to say, the elements of A are subsets of X.' For example if $A = \{A_1, A_2, \ldots\}$ and $X = \{X_1, X_2, \ldots\}$, we may have $A_1 = \{X, X_3\}, A_2 = \{X_2, X_4, X_5\}, \ldots$. The elements of A need not be mutually exclusive subsets of X — overlap is permitted.

If the set A is taken as defining some system at the $(N + 1)$-level of resolution, then the elements of X may be taken as the components of that system at the N-level. This carries connotations of functional relationships and behaviour patterns of A which are not immediately obvious from seeing a list of the elements of A — a point which should be stressed in a systems theory context, but which is not stressed by Atkin. This then leads to a comment about hierarchies in systems: levels of resolution must be carefully specified, and objects in different levels properly distinguished as such. But when this is done, much of the workings of the system will be evident from the hierarchical organization thus revealed. This point will be taken up again in Chapter 9 in the context of hierarchical control systems.

The relation between the elements of sets such as A and of X can be expressed as an incidence matrix μ_{ij}: $\mu_{ij} = 1$ if A_i contains X_j, and zero otherwise. It is then possible to carry out q-analysis on such a relation using the methods described earlier. We should emphasize, however, that relations such as λ used in Section 7.2.8 were at one hierarchical level, say between the elements of the sets X and Y in Figure 7.37. Clearly this shows up another kind of hierarchical structure: a simplex with a higher q-value than another is in some sense a higher order — so this is another kind of hierarchy within a single level of a hierarchy determined by levels of resolution.

Finally, we recall the use of slicing parameters to identify binary relationships for q-analysis from, say, flow data which relates two sets. Different slicing parameters will then, in effect, identify different hierarchical structures in the data. It would almost certainly be possible, for example, to show that Nystuen's and Dacey's method was a special case of such a technique.

7.2.10 Networks and decisions

There are two techniques which represent parts of network analysis which are best illustrated in relation to management and decision making processes. The first is critical path analysis, which is mainly used for job scheduling when there are large numbers of component tasks related in a complicated way; it also turns out to give interesting new insights for systems theory. The second is the so-called *branch-and-bound* method, which is an approach to solving integer programming problems and so in a sense belongs in Chapter 6; however, it can be presented as a decision procedure and involves a network and so also conveniently fits here.

Let J be a set of jobs $\{J_1, J_2, J_3, \ldots\}$; let T be a set of times which it takes to complete each job — $\{t_1, t_2, t_3, \ldots\}$. Each job can be considered to start with an initial event, labelled by an index i, and a final event, labelled by an index j. Let $\mathbf{T}$ be a set of events labelled $\{1, 2, \ldots, i, j, \ldots, N\}$. Then each job, j_r, has a pair of events associated with it, say (i_r, j_r). The interdependence of jobs is then represented by

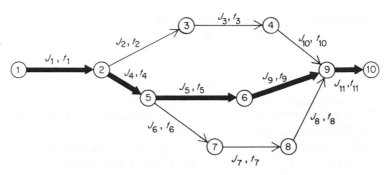

Figure 7.38 Jobs and times for a critical path analysis

the sharing of event labels. For example, if job J_3 is an input to job J_5, and there are no intermediate jobs, then $j_3 = i_5$. This, of course, defines a network, such as that shown in Figure 7.38. The time associated with each job can then be assigned to the network link, connecting two events, which represents the job. The data are also presented in Table 7.11.

There are three paths in this network connecting events 1 and 10: $\{1, 2, 3, 4, 9, 10\}$, $\{1, 2, 5, 6, 9, 10\}$, and $\{1, 2, 5, 7, 8, 9, 10\}$. The total times on each path can easily be seen to be 20, 41, and 30 respectively. The second one, shown by the double line in Figure 7.38, is known as the *critical path*, since it identifies the minimum amount of time needed to complete the whole job and also the amount of 'spare time' in the other paths. This could lead to a revised allocation of resources to jobs. In the example, any method of speeding up job J_5 can be seen to be particularly valuable.

This is an obviously useful technique in itself in a wide variety of circumstances. Its wider use in systems theory arises as follows. The behaviour of any system is governed by the behaviour of its components and the interdependencies between them. Since component 'jobs' take time, and the interdependencies mean that they receive inputs from other components, the whole process for particular

Table 7.11

Job J_r	Initial event label, i_r	Final event label, j_r	Time	Example time
J_1	1	2	t_1	10
J_2	2	3	t_2	2
J_3	3	4	t_3	3
J_4	2	5	t_4	3
J_5	5	6	t_5	20
J_6	5	7	t_6	3
J_7	7	8	t_7	5
J_8	8	9	t_8	6
J_9	6	9	t_9	5
J_{10}	4	9	t_{10}	2
J_{11}	9	10	t_{11}	3

'behaviour' sequences could in principle be represented by something like a critical path network. This would determine the time for the whole process and the amounts of 'leeway' (or 'redundancy') in other paths in that network. Thus, as an analytical method, we might expect this approach to be useful in the study of ecosystems. For a city, it would be an interesting basis for the planning of complicated projects — for example, the phased build up of a new town where it would be necessary to keep houses, jobs, and services in balance.

The second technique to be dealt with here is the 'branch and bound' method in the form presented by Scott (1971). This is concerned with an optimization problem of the form

$$\text{Minimize } Z = f(x_1, x_2, \ldots, x_N) \tag{7.39}$$

subject to

$$g_i(x_1, x_2, \ldots, x_N) > b_i \qquad i = 1, \ldots, M \tag{7.40}$$

$$x_i = 0 \text{ or } 1 \qquad i = 1, 2, \ldots, N \tag{7.41}$$

where f and g_i are monotonic *increasing* functions of the variables $x_1, x_2, \ldots, x_N$ which are themselves integers restricted to the values 0 or 1.

This is quite a common decision problem, for example in the allocation of service facilities. The objective function may represent cost minimization and the constraints various conditions — schools must be available for all children of appropriate age, and so on. The problem can be represented in a tree-like network as in Figure 7.39. All variables $\{x_i\}$ are initially set to zero; the next generation of nodes each have one x_i set to 1 — denoted by (x_i) — then pairs are set to 1, and so on.

At each generation Z is calculated for each node and the constraint set is

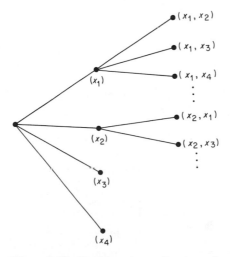

Figure 7.39 Decision trees for branch and bound

evaluated. The lower bound on Z, L, is the minimum value of Z yet achieved (whether the constraints are satisfied or not) and the upper bound, U, is the best *feasible* value. The power of the method is that, once a node has a value of Z which is bigger than the current value of U, no further examination of that branch of the network is necessary (since the monotonicity conditions mean that Z will continue to increase down that branch from a value which is already worse than the best which has been achieved already). Another effect of monotonicity is that as the calculation proceeds, L increases, and the procedure terminates when $L = U$, and this is the optimum value of Z.

7.2.11 Network change: design and growth

So far, the techniques outlined for different aspects of network analysis have been concerned with existing, static, networks. Now, we turn to the study of network change and evolution. This is a case where it is important to bear in mind the distinction between analysis (of change) and planning, and, in the former, whether planners may have directly influenced the form of evolution or not. In the purest form of analysis, it may be assumed that 'forces', to be identified, drive the form of network evolution; in a less pure form, one at least of these forces may be the planning–governmental system of the time. In a pure planning problem, it will be assumed that designs are sought to optimize some criteria — usually by adding to an existing network. Thus, first we discuss approaches to the study of growth, and then the design problem as an optimization problem.

Haggett and Chorley (1969) make a useful distinction between network growth as 'node connecting' and as 'space filling'. (They also add 'space partitioning', but this is less relevant to our immediate objectives.) In the first case an existing set of

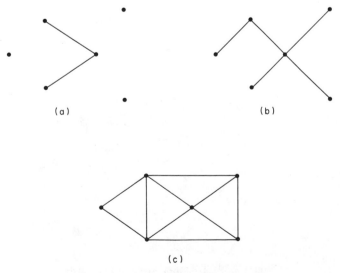

Figure 7.40 'Space-filling' network growth

nodes is assumed and links are progressively added, as shown in Figure 7.40. This may be a good model for the development of an inter-urban network where a large set of relatively isolated nodes would exist at a time of poor transport technology and economic development, and links would be added progressively over time. It would be easy to establish which were likely to be the most 'important' links, and which would therefore come first, and a version of this method has indeed been used by MacKinnon and Hodgson (1969—cited by Haggett, Cliff, and Frey, 1977).

The space-filling approach is more appropriate to something like river systems and geomorphic processes. Here, the givens are areas rather than nodes, and while the landforms appear to determine the structure of the river system, over the long term the reverse is more likely to have been true. A key question for geomorphologists is therefore of the development of river networks and the feedback-impact of this process on landforms. One method which has been used for explaining such processes is simulation modelling (beginning with Langbein and Leopold, 1964). This can be accomplished by beginning with a square grid as shown in Figure 7.41. Some first-order streams are added to initiate the process and then proceed on a probabilistic random walk basis as shown in subsequent parts of Figure 7.41. When two first-order streams meet in one square, they are then considered to form a second-order stream, and so on.

One of the interesting consequences of this approach is that it has generated connections to entropy maximizing methods. If a large number of simulation model runs are carried out, certain 'average' proportions of the resulting networks, such as branching ratios, can be calculated and it turns out that these

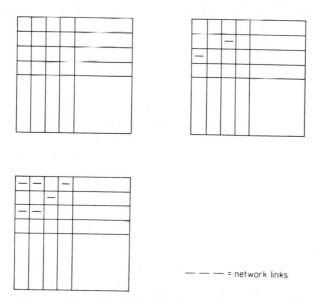

— — — — = network links

Figure 7.41 A grid as the basis for simulating network growth

are quite stable and in accord with earlier work. The procedure involves the identification of so-called 'topologically distinct channel networks', and the computation of these — which is a combinatorial process — and the calculation of higher scale average quantities is essentially an entropy maximizing method of the type outlined in Chapter 4.

It is perhaps appropriate, in the light of the first part of this subsection, to add a third category to those of Haggett and Chorley which may be called 'hierarchical evolution'. We have seen that links of networks can be of different orders in a hierarchy and have discussed the 'trunking' process. This clearly adds an extra dimension to the study of network evolution. For example, though pairs of nodes can be connected, as discussed earlier, there will be later stages of development where low-capacity links added at an early stage will be replaced by higher-order links.

Finally, we turn to network optimization problems — where growth can be planned to satisfy some objective function. Such problems are usually very difficult for realistic networks because of their combinatorial nature and the large number of nodes and links involved. Network optimization has two main features: whether to add links or not, and the specification of the size of each link. Steenbrink (1974) notes that, for n possible links, each of m possible dimensions, the total number of combinations is m^n. For example, if there are 30 possible links and two dimensions (present or not), there are 2^{30} — over a thousand million — possible combinations. In the Dutch work reported by the same author, the case study involved 351 origin and destination nodes, 2000 other network nodes, and 6000 possible links. The initial statement of the optimization problem then involves approximately 750 million variables and 250 million constraints!

Network optimization problems can be stated in a variety of ways and a few of the better known and more useful ones will be described here. Perhaps the simplest interesting problem is that of the *minimum spanning tree*: given n nodes, how to allocate $n-1$ links to connect all the nodes at minimum cost. The kind of network involved is exhibited in Figure 7.42. It is interesting to observe, following Haggett, Cliff, and Frey (1977) that such a tree is 'a network of links and nodes such that there is exactly *one path* connecting any two nodes' (my italics) — and this gives a new insight into the nature of tree networks, which were defined in an earlier section as being without circuits. A formal statement of the problem is

$$\operatorname*{Min}_{\{c_{rs}\}} Z = \sum_{rs} \lambda_{rs} K_{rs}$$

$$\sum_{rs} \lambda_{rs} = n - 1 \tag{7.42}$$

where the K_{rs}'s are link building costs and λ_{rs} is one or zero according to whether a link is built to connect r and s or not. The algorithm for solving the problem, as stated by Haggett, Cliff, and Frey, is first to connect each node to its nearest neighbour (in this case presumably assuming that the unit defining 'nearest' is

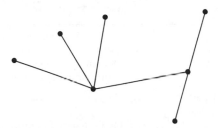

Figure 7.42 A minimum-spanning tree

directly proportional to what we have called building costs), then to connect each subnetwork so formed so its nearest neighbour, and so on until the minimum spanning tree is complete.

The precise nature of the problem, and the methods to be used to solve it, depend on the precise formulation of the objective function and the constraints. The minimum spanning tree problem, used as an initial example above, makes no reference to any demand for the *use* of links. An obvious next step, therefore, would be to seek an optimum network which took into account say a fixed origin–destination trip matrix. Such a step, as will be evident from much of the preceding sections of this chapter, immediately complicates the problem considerably: the origin–destination flows have to be assigned to 'best' paths and link loadings and link costs related to these. The next complication to be added is to make the origin–destination flows dependent on best-path costs, which are related to flow-dependent link costs as described in Section 7.2.5. In order to make all these relationships explicit, a variable array such as $\{x_{ij}^{rs}\}$ must be used: this is the flow from i to j which is part of the load on link (r, s)— and this is why Steenbrink gets such enormous numbers of variables in his example, essentially arising from this four-dimensional array.

The final aspect of problem definition is the objective function. The essential distinction is that implied by Wardrop's two principles as discussed in Section 7.2.4: whether it is assumed that overall system costs are being minimized and that people or goods can be assigned to the system, or the more realistic assumption that individuals optimize their own costs (in effect playing a game against all others in the system until equilibrium is achieved) while the government minimizes its own costs subject to this individual behaviour as a constraint.

The types of network optimization problem sketched above are summarized in Table 7.12. This table defines 6 possible problems.

There are essentially three broad approaches to seeking optimal networks, of which the minimum spanning tree can be the initial stage of the first: that is, to start with a minimum set of links and add to this set until the budget is spent— seeking to optimize an objective function at each stage. The second approach is to begin with a complete listing of possible links and then progressively to delete links until the constraints are satisfied and the objective function optimized. The

Table 7.12

Objective function	System definition
Optimal total system cost	Demand not taken into account (minimum spanning tree)
Individual optimization plus government objective function	Fixed O–D matrix
	O–D matrix function of costs

third approach involves mathematical programming. As can be seen from the discussion in Section 7.2 on branch and bound methods, the first two approaches essentially fall into this category. As links are added, for example, the objective function moves nearer to its optimum value and checks can be made as to whether the constraints are satisfied. Such methods, in the network context, are described by Boyce, Fahri, and Weischedel (1974). It turns out that the results of the minimum spanning tree often provide good starting points for such algorithms, but of course it is clear that each of the first $n-1$ links will not *necessarily* be part of the final solution which may lie down another branch of the decision tree.

Programming methods of solution exist in theory for the kinds of problems outlined—the problem essentially being that of Section 7.2.5 with the network links added as variables—but there is the enormous practical difficulty of size of problem and this is where solution efforts have been concentrated. The reader is referred to Steenbrink (1974) for a discussion of various possible methods. The one used by Steenbrink is a decomposition method, essentially involving an embedding of nested programming problem of the type discussed in another context in Chapter 6. He also discusses the possibility of a hierarchical structuring of the problem—first seeking the highest order links, then the next level, and so on. The methods of Boyce, Fahri, and Weischedel, and of Steenbrink, have been applied in practice, but not widely, and this is a field of research we would expect to see expand in the future.

A final note relates to the analysis–planning distinction raised at the beginning of this subsection. The discussion on optimization has been set in a planning context, but if it is accepted that the 'forces' implied by these procedures were taken into account, intuitively, by earlier generations, then optimization methods can be used analytically in explorations of network growth. A number of such analyses, based on minimum spanning tree algorithms, are described by Haggett, Cliff, and Frey (1977).

7.3 CONCLUDING COMMENTS

In Section 7.1, we outlined the importance of networks in systems analysis and illustrated some different network types. But the heart of the chapter has been

Section 7.2 in which we have sketched a wide range of techniques. These exhibit some of the usual characteristics of methods of systems analysis, and in particular that they vary from the practical and useful (and easily intelligible!) to those which address complex and difficult theoretical problems. It is frequently necessary to find the shortest path through a network, but also to relate this to flow assignment and flow–link–time interdependence. It was also interesting to see how these concepts tied up with mathematical programming in Section 7.2.5, thus establishing connections to Chapter 6.

The following three subsections, on connectedness, address a similar problem in different ways. Different measures of connectedness seem, intuitively, to be important in tackling complex systems—but they are not easily built directly into models. It may be, therefore, that these techniques will be used on their own account for the present, but that they represent the beginnings of a substantial research programme which will involve building them directly into models—rather as in the case of shortest path or equilibrium methods.

The final three subsections are also concerned with difficult problems. The most straightforward is that on decisions, which gives some routine procedures for problems of certain types. The problem of hierarchies is one which recurs in various ways throughout the book, and we saw in Section 7.2.9 that network analysis offers some interesting perspectives. Finally, in Section 7.2.11, we tackle the problem of network evolution. Here we concentrate on reviewing a number of approaches, but a further specific model will be presented in the context of dynamical methods in Chapter 8.

In summary, therefore, we can note the existence of a number of techniques which are fundamental to many system models, and the existence of a number of approaches to some difficult questions—like connectedness, hierarchical structure, and evolution—which form the basis of some important research problems. In some of these cases, research needs to be carried out on the basis of particular case studies, and one of the hopes of the author in presenting these ideas here is that students may be able to contribute to them in undergraduate or postgraduate dissertations.

NOTES ON FURTHER READING

Useful background for geographers is provided by †Haggett and Chorley (1969). There is also a useful geographical treatment of networks in a transport context by †Hay (1973, especially Chapters 3–6) and a geographical-transport case study is offered by *O'Sullivan, Holtzclaw, and Barber (1979). *Potts and Oliver (1972) provide a formal treatment of many aspects of networks in a transport context. †Scott (1971, Chapters 5 and 6) provides useful material on a number of aspects of network analysis.

A number of other references provide useful background reading for particular subsections. An integration of spatial interaction models and network flow equilibrium under a mathematical programming umbrella is provided by *Evans

(1976) (7.2.5). An extensive treatment of digraphs and related concepts is offered by *Roberts (1976) (7.2.7). Q-analysis is described in two books by *Atkin (1974, 1977). Hierarchical structure in network is interestingly tackled by *Woldenberg (1970) (7.2.9). There is an elementary treatment of 'branch and bound' in †Scott (1971, Chapter 2) and a more detailed one in *Boyce, Fahri, and Weischadel (1973) (7.2.10). *Steenbrink (1974) describes some large network design problems (7.2.11).

CHAPTER 8

Dynamics

8.1 APPROACHES TO DYNAMICAL SYSTEMS THEORY

8.1.1 Introduction

Dynamical systems theory is, of course, concerned with the way in which some systems of interest changes through time. A number of foundations for the study of change have already been laid in previous chapters in relation to a number of basic concepts of systems theory: the representation of systems, including the definition of state variables and phase space; comparative static approaches, including steady state; the accounting foundations; optimization processes producing equilibrium states and thus facilitating a particular comparative static approach. We will return to these aspects shortly and emphasize the role of these methods in the study of change (as distinct from 'state at a *point* in time' which has been the emphasis of previous chapters). We will then go on to explore new methods of dynamical systems theory.

A good starting point is the usual one of identifying the state variables of some system of interest at a point in time. The exogenous variables of such a system can be considered as the inputs and the endogenous variables as the outputs. The system may be a subsystem of some wider system, so that the inputs may be outputs of other subsystems, and vice versa. This is sketched in Figure 8.1. The possibility of feedback mechanisms is included. Even *parameters* (say representing structural features often considered 'fixed' in the short term) of the system of

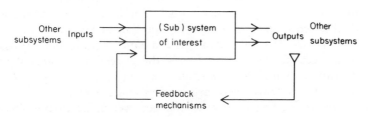

Figure 8.1 Feedbacks

179

interest can be considered as exogenous *variables* in this representation as some of them at least can in principle be changed.

A dynamic model will consist of rules for calculating endogenous variables for given values of exogenous variables (considered given as functions of time). The latter, therefore, together with the interval structure, or rules, can be considered as the *driving force* of the system of interest. If these distinctions are made carefully, dynamical analysis is much easier. In the discussion of methods which follows, based on these foundations, as many simplifying assumptions as possible will be made so that the dynamical systems' aspects are emphasized. For example, as we saw in the previous Chapter, many system models are inherently complicated because of congestion on underlying networks; we will mostly ignore such complications in this chapter. We will also tend to assume that variables and mechanisms are *deterministic* rather than stochastic (which is 'probabilistic'). There are various reasons for this: first, it is often a good approximation anyway; secondly even with stochastic variables, much useful modelling can be conducted with their means; and thirdly, explicit stochastic modelling would take us beyond the mathematical techniques available in this book. The main methods of dynamical systems theory will now be sketched in turn and then considered in more detail, with examples, in subsequent sections.

8.1.2 Comparative statics

The methods of 'comparative statics' usually involve an assumption that the system of interest is governed by a mechanism which maintains it in an *equilibrium* state, and that the system returns rapidly to a possibly new, equilibrium state after it has been disturbed. Systems involving a *steady state* fall into the same category: although there is motion in such systems, the flow variables take constant values, and these values are usually maintained by some equilibrium mechanism.

Thus comparative static modelling methods are, more or less, those of Chapters 4, 5, and 6. Time will not appear explicitly in model equations. If any exogenous variable is changed, then the value of the endogenous variables at a new equilibrium position can be calculated. Nothing is said about the *transition* to the new equilibrium so that those methods are usually only applicable if the *relaxation time* of the system is short. Otherwise, the system's *transient states* have to be modelled and this involves some of the other methods to be discussed below.

The complexities which can be represented by comparative static methods should not be underestimated. This is because models of systems made up of connected subsystems can represent very complicated behaviour indeed. This can arise even when most of the subsystems are comparative static. It happens for two reasons: first, because the driving forces, the exogenous inputs of different subsystems, can be changing at different rates; and secondly, because even in comparative static systems, there are circumstances where small changes in exogenous inputs can bring about large changes, even in fact *discrete* changes, in endogenous outputs. (And recall that these outputs may be other subsystem inputs, and so on.)

8.1.3 Difference equations

Consider accounting array totals such as $K^{i*}(t)$ and $K^{*i}(t + T)$, equal, say, to $P^i(t)$ and $P^i(t + T)$ respectively. Then the difference in type-i numbers at times $t + T$ and t is obviously $P^i(t + T) - P^i(t)$. A δ or Δ sign is sometimes used to denote this:

$$\delta P^i(t) = P^i(t + T) - P^i(t) \tag{8.1}$$

or

$$\Delta P^i(t) = P^i(t + T) - P^i(t). \tag{8.2}$$

Note that the time interval, T, is implicit in, and important to, the definition. This itself is sometimes written as δt or Δt rather than T. For those unfamiliar with this notation, it should be emphasized that δ (or Δ) is a symbol standing for 'increment in whatever follows'; it is *not* in itself an algebraic variable though δt or Δt is. Life is made all the more complicated because δ and Δ sometimes are used as algebraic variables, though this should always be clear in a particular context and such uses will usually have nothing to do with time increments.

We can now recall the accounting equation (5.129) from Section 5.6. This implies

$$K^{*i}(t + T) - K^{i*}(t) = \Delta P^i(t) = \sum_{i \neq j} (K^{ji} - K^{ij}) \tag{8.3}$$

where $\{K^{ij}\}$ is the unerlying accounting array. If we defined row sum coefficients, a_{ij}, such that

$$a_{ij} = K^{ij}/K^{i*} \tag{8.4}$$

then (8.3) can be written

$$\Delta P^i(t) = \sum_{i \neq j} [a^{ji} P^j(t) - a^{ij} P^i(t)]. \tag{8.5}$$

This is a set of equations in the variables $P^i(t)$ and difference terms, $\Delta P^i(t)$, and they are known as difference equations.

Sometimes it is useful to define differences of differences. Second order differences could be defined by

$$\Delta^{(2)} P^i(t) = \Delta^{(1)} P^i(t + T) - \Delta^{(1)} P^i(t) \tag{8.6}$$

using an obvious notation. If differences of this type appear in the equations, then such difference equations are said to be second order.

8.1.4 Differential equations

Problems stated in terms of difference equations can always be stated in terms of differential equations and vice versa. The former are used when the time interval, $T, \delta t$, or Δt, according to the notation used, is finite; the later when the time interval is infinitely small—an infinitesimal. So differential equations are constructed from difference equations by letting the time increment tend to zero. We now show the notation for this. The average rate of change of the variable given in equation (8.7) in a period δt (where we have switched notation on increments for convenience) is

$$\frac{\delta P^i(t)}{\delta t} = \frac{P^i(t + \delta t) - P^i(t)}{\delta t}. \tag{8.7}$$

If $P^i(t)$ is plotted against t, this average is shown as the gradient of the chord AB in Figure 8.2. $\delta P^i(t)/\delta t$ is BC/AC as shown in the figure. What happens as $\delta t \to 0$? The point B moves towards A down the curve, and eventually the chord AB becomes the *tangent* to the curve at A—shown by the dashed line in the figure. The gradient of the tangent is the instantaneous rate of change. It is written $dP^i(t)/dt$ and the notation for the limit is

$$\frac{dP^i(t)}{dt} = \mathop{\mathrm{Lt}}_{\delta t \to 0} \frac{\delta P^i(t)}{\delta t}. \tag{8.8}$$

This is known as the *derivative of* $P^i(t)$. The reader will have noticed that in the fraction on the right-hand side of (8.7), both numerator and denominator tend to zero as $\delta t \to 0$. Nonetheless, techniques exist for calculating the actual value of this limit—the methods of the differential calculus. For the purposes of this book, it is simply necessary for the reader to know that the gradient can be calculated and that equations can be stated which involve gradients—differential equations, which apart from the fact that they involve derivatives (or *differentials*) rather than finite differences are very similar.

Consider some variable x. A simple differential equation is

$$\frac{dx}{dt} = ax \tag{8.9}$$

where a is a constant. That is, the instantaneous rate of growth of x is proportional to x itself. This is the basic equation of exponential population growth. Its solution is

$$x = e^{at}. \tag{8.10}$$

A dot over a variable is sometimes used to represent differentiation with respect to time, so that (8.9) could be written

$$\dot{x} = ax. \tag{8.11}$$

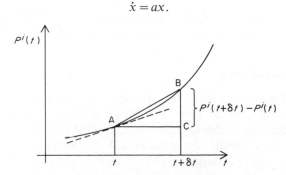

Figure 8.2   The gradient of a curve at a point

The difference and differential equations considered so far all involve constant coefficients and no interaction between state variables. However, such interactions can occur as in the pair of equations

$$\dot{x}_1 = (a - bx_2)x_1 \qquad (8.12)$$

$$\dot{x}_2 = (cx_1 - d)x_2. \qquad (8.13)$$

If we compare (8.11) with (8.12), we see that a has become $(a - bx_2)$ which can be interpreted as a constant rate of growth being reduced by bx_2. This represents an interaction, or in Chapter 2 terms, a *coupling*, between x_1 and x_2. Similarly, there is a coupling term, cx_1x_2, in equation (8.13). Thus we begin to see that differential equations can be useful in describing system behaviour. This particular example will be pursued further in Section 8.4 below.

8.1.5 Dynamical systems theory from a topological viewpoint

The starting point for a topological viewpoint is the idea of a function as a *mapping* (with the term being used very much in a mathematical sense rather than a geographical one). Thus, if y is a function of x, and we write

$$y = f(x) \qquad (8.14)$$

This function represents a mapping of x to y; given a value of x, it is a *rule* for getting a corresponding value (or, occasionally, values) of y. The main point of using what at first sight seems to be a rather elaborate definition of a function is that it allows more difficult problems to be solved (or at least new insights to be gained) than traditional methods.

In the case of dynamical system theory, the function, or mapping, is, in effect, the set of model equations which relate the endogenous (state) variables to the exogenous variables—say relating a vector of state variables $\mathbf{x}$ to a vector of inputs, $\mathbf{u}$. Formally, we can write:

$$\mathbf{x} = \mathbf{f}(\mathbf{u}) \qquad (8.15)$$

which is

$$x_1 = f_1(u_1, u_2, \ldots, u_M)$$
$$x_2 = f_2(u_1, u_2, \ldots, u_M)$$
$$\vdots \qquad (8.16)$$
$$x_N = f_N(u_1, u_2, \ldots, u_M).$$

The possible values of $\mathbf{x}$ and $\mathbf{u}$ can be considered to form a surface, or *manifold*, in a high-dimensional space (in fact of dimension $M + N$). The behaviour of a dynamical system can then be seen as a path *on that manifold*. Thus, dynamical systems theory from a topological viewpoint is concerned with the nature of such manifolds and possible behaviours upon them. One way in which manifolds can be generated is through sets of *constraint* equations such as those which played a major role in Chapters 4 and 5 and whose general role was described in Section 6.3. Further, the distinction between comparative static and other approaches fades somewhat because time can be treated as 'just another variable' and an $(M + N + 1)$-dimensional manifold investigated.

There is a sense in which the mathematics involved becomes qualitative. *Generic* properties of types of system are sought rather than detailed numerical results (though these are sometimes achievable also). An example of the kind of phenomena which can be handled under this representation but not using the traditional method of calculus is *discrete* change in systems. In particular, this involves catastrophe theory, which will be explored further in Section 8.5 but also turns up in the modern theory of differential equations in one of the examples of Section 8.4.

8.2. COMPARATIVE STATICS AND MATHEMATICAL PROGRAMMING

8.2.1 Power in shopping facility planning

To find interesting 'comparative static' behaviour in a mathematical programming context, we return to the example introduced in Section 4.6 of a government agency optimizing the distribution of shopping centres to maximize consumers' surplus, knowing that the consumers' behaviour could be described by an entropy maximizing model. Now, however, we assume that the government itself has an objective—to weight central facilities (described by $j \in C$, where C is the set of central zones) against suburban facilities ($j \in S$) in the ratio $1 : \rho$, and that producers seek to maximize their profits. Suppose such profits are proportional to turnover in a zone raised to some power, θ, to represent the profitability of large centres, less some fixed cost. So producers' profit at j is

$$P_j = \gamma(\sum_i S_{ij})^\theta - W_j p_j \tag{8.17}$$

say. We can now state the overall problem as

$$\operatorname*{Max}_{\{S_{ij}, W_j\}} Z = \lambda\left[\sum_{j \in C} W_j + \rho \sum_{j \in S} W_j\right] + \mu\left[-\frac{1}{\beta}\sum_{ij} S_{ij}\log(S_{ij}/W_j^\alpha) - \sum_{ij} S_{ij} c_{ij}\right]$$
$$+ \nu\left[\sum_j \{\gamma(\sum_i S_{ij})^\theta - W_j p_j\}\right] \tag{8.18}$$

such that

$$\sum_{j \in C} W_j > W^C \quad \text{(government)} \tag{8.19}$$

$$\sum_j S_{ij} = e_i P_i \quad \text{(consumers)} \tag{8.20}$$

$$W_j > W^{\min} \text{ (producers)}. \tag{8.21}$$

λ, μ, and ν are weights which represent the 'power' of the different agents

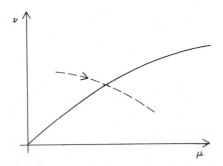

Figure 8.3 A curve of critical parameter values

represented in the objective function; and we have also shown typical constraints for each agent: the government want minimum size in the centre, consumers to spend all their money, and producers to have a minimum size in a zone.

This mathematical programme has a solution at any point in time. The interest from a dynamic point of view arises from considering what happens to this solution when λ, μ, and ν change, relatively to each other, over time. The possibility exists of a sudden and dramatic change in one or more of the W_j's for a small change in $\lambda:\mu:\nu$. To make this definite, take $\lambda = 1$ (as there are only two independent variables here) and let μ and ν vary. Then, there may be a curve in the (μ, ν) plane such that, if the (μ, ν) trajectory crosses this curve, a sudden change in the W_j's follows. This is exhibited (purely hypothetically) in Figure 8.3. The dashed line is the (μ, ν) trajectory: in effect, it shows consumers becoming more powerful relative to producers. The continuous curve is such that, if it is crossed, the W_j's change discretely. Such a curve is called a *bifurcation curve* and the area near it is a *critical region* in (μ, ν) space.

The discrete changes could easily arise in this particular case from the optimum solution of the mathematical programme jumping to a new corner of the feasible region defined by the constraints as the weights in the objective function change. This possibility can also be related to Hotelling's famous ice-cream man on a linear beach problem presented in Figure 8.4. This problem is well known and

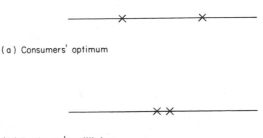

(a) Consumers' optimum

(b) Producers' equilibrium

Figure 8.4 The Hotelling 'ice-cream-man-on-a-li-
near-beach' problem

will be sketched very briefly. There is an even density of people on the beach and two ice cream men to locate. The consumers' optimum is obviously to have the ice-cream men at the quartiles as shown in Figure 8.4(a). However, if producers seek to maximize their profits, A (first, say) moves towards the centre to try to capture more of the market; the only way B can compete is to do likewise. And so the stable equilibrium when producers' profit describes the system is Figure 8.4(b) with both men at the centre. They still each get half of the market, but consumers travel further for their ice cream. It seems a reasonable conjecture that Figure 8.4(a) represents a (large μ, small v) solution and Figure 8.4(b) a (small μ, large v) one. Thus the problem discussed above may be a more complex N-facility, unevenly distributed population, version of a similar problem.

This argument alerts us to the possible existence of *critical regions* in the state space of the *exogenous* variables (λ, μ, and v in this case). It remains a major research task to identify such regions, though the importance of this for various aspects of planning is clear.

8.2.2 Trunk routes in networks and related problems

We next pick up a problem which was first considered in the previous chapter: how do trunk links and routes (or more generally higher-order links and routes) come about in networks? Woldenburg suggested that some kind of least work principle may be operating in natural networks such as river systems or trees. Here, we consider a nonlinear mathematical programming model of a simplified problem which could offer new insights.

Suppose some facilities exist in discrete size groups. Let X_j be the size of the jth class, and n_j the number of facilities of this type. We can assume that both the cost, $f(X_j)$, and the capacity, $F(X_j)$, of the facility increase with size but that scale economies give these functions the shapes shown in Figure 8.5. Then the mathematical programming problem will involve minimizing cost *and* providing sufficient capacity. The constraints on capacity may take two forms: first, a minimum number of facilities (of any size), N, must be provided; and secondly an

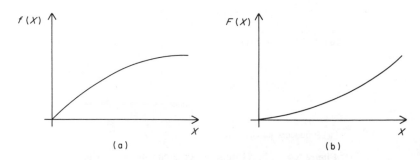

Figure 8.5 (a) Facility cost by size; (b) capacity by size

overall capacity, C, must be met. Thus the programme is

$$\operatorname*{Min}_{\{n_j\}} Z = \sum_j n_j f(X_j) \qquad (8.22)$$

such that

$$\sum_j n_j > N \qquad (8.23)$$

and

$$\sum_j n_j F(X_j) > C. \qquad (8.24)$$

It is clear that, for small c, a lot of small facilities will minimize cost (in fact, $n_1 = N, n_j = 0, j > 1$ for very small c), but that as c increases, higher-order facilities will be required. We thus have a model of the evolution and genesis of higher order facilities.

So far, we have not interpreted the model. Possible examples are shops of different orders with c representing overall consumer expenditure and N generating a minimum number of shops per square mile. Another example would be the provision of roads of different order in a corridor: C would be overall capacity while N would generate a minimum distance to gain access to at least a low-order road. A third example is the number of streams of different orders in a river network. The cost functions could be reinterpreted as 'work functions' (for carrying a unit volume of water), N would generate a minimum density of streams for carrying the water in a particular basis, and C would represent overall basin runoff in a period. A similar analysis could be applied to a tree network, with N generating a high density of branches (to provide access to sunlight for the maximum number of leaves) and C the capacity needed for carrying nutrients from the soil and for providing a suitably strong structure for the tree. Finally, j may be taken to label plants of different sizes in an ecosystem. N would again guarantee taking maximum advantage of sunlight, while C would be a measure, say, of the nutrient capacity of the soil.

8.3 DIFFERENCE EQUATIONS

8.3.1 Introduction and an example

We saw in Section 8.1.3 that difference equations relate variables such as $x_i(t)$ and their differences $\Delta x_i(t) = x_i(t + T) - x_i(t)$, say. We also noted that difference equations can be converted into differential equations and vice versa. Indeed, differential equations which are to be solved numerically must be converted into difference equations for computational purposes. Most of our analytical interest can be concentrated on differential equations and discussion postponed to the next section. However, one major development based on difference equations is Forrester's 'system dynamics' and this can be used to illustrate the method and the way it can be deployed computationally.

We use a slightly modified version of Forrester's notation, to be more in accord with this book, but otherwise follow Figure 8.6 which is taken from Forrester's

188

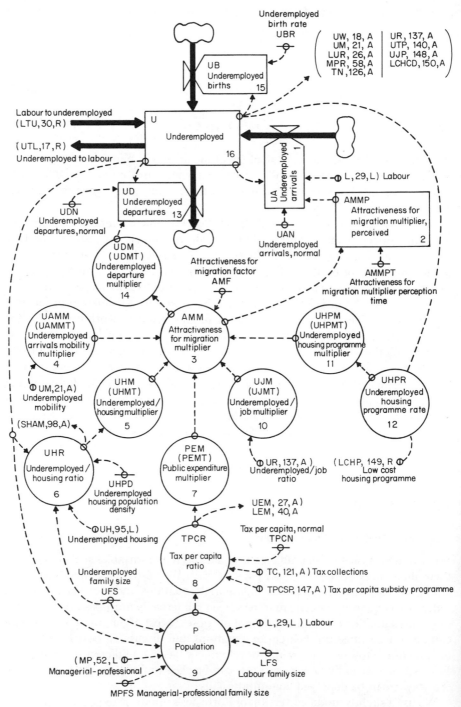

Figure 8.6 Part of a system from Forrester's *Urban dynamics*

(1969) *Urban dynamics*. The task is to calculate the value of the land variable, U, for a time $t + T$. It is given by an accounting equation

$$U(t + T) = U(t) + T(UA(t, t + T) + UB(t, t + T)$$
$$+ LTU(t, t + T) - UD(t, t + T) - UTL(t, t + T)) \qquad (8.25)$$

where UA, UB, LTU, UD, and UTL are various (positive and negative) elements of change in the accounts. $UA(t, t + T)$, to follow through as an example the treatment of one of these variables, is given by

$$UA(t, t + T) = (U(t) + L(t))(UAN)(AMMP(t)) \qquad (8.26)$$

where UAN is a parameter, $L(t)$ is the labour force (so $U(t) + L(t)$ is the total potential workforce) and $AMMP(t)$ is a measure of the perceived attractiveness for migration. This in turn is a product of a number of factors:

$$AMMP(t) = AMMP(t - 1) + T(AMM(t - 1)$$
$$- AMMP(t - 1))/AMMPT. \qquad (8.27)$$

$AMM(t)$ is then given by another equation, and so on through all the connections implied by Figure 8.6.

Thus, the method is simple to apply: set up an equation for the next variable down a chain in terms of variables at the next lower level. The overall behaviour of the model can be very complicated because of various feedbacks (which are clearly visible in Figure 8.6 for example). The main drawback, however, in building up difference equation models in this way is that each particular equation, as can be seen, usually involves a very simple assumption about change in that independent variable and this may often be inadequate for modelling many systems.

8.3.2 Complicated dynamics from simple equations: applications to ecosystems and cities

May (1976, for example) has drawn attention to a number of results involving very simple difference equations where the possible modes of behaviour are very complicated. His work was developed in an ecosystems' context but has applications in a number of other fields and turns out to be directly applicable, after one modification to generate a difference equation from a differential equation, to shopping centre dynamics.

Indeed, it is appropriate to start with a well known differential equation. Consider the equation

$$\frac{dx}{dt} = \dot{x} = ax(1 - x/b). \qquad (8.28)$$

This is rather like equation (8.11) of Section 8.1.4 but with the term $1 - x/b$ added. This term inhibits the exponential growth of the simpler equation by a mechanism which can be considered as negative feedback. As x grows, $1 - x/b$

declines and 'reduces' the effect of the parameter a until at $x = b, 1 - x/b$ is zero, and hence so is x. Thus both $x = 0$ and $x = b$ are equilibrium points of (8.28); $x = 0$ is unstable, $x = b$ is stable. This is in fact the equation of logistic growth. If x is initially some small quantity, much less than b, then the solution of the equation, for x against t, is the familiar S-shaped growth curve to an asymptotic upper bound.

May's example is based on the difference equation version of this process. As a finite approximation, we could take

$$\dot{x} = \frac{\delta x}{\delta t} = \frac{x_{t+\delta t} - x_t}{\delta t}. \tag{8.29}$$

For convenience, and without loss of generality, we can take $\delta t = 1$. Thus

$$\dot{x} = x_{t+1} - x_t \tag{8.30}$$

and the difference equation version of equation (8.28) is thus

$$x_{t+1} - x_t = ax_t(1 - x_t/b) \tag{8.31}$$

(where subscripts t have been added to the x-variables on the right-hand side). This can now be written

$$x_{t+1} = (1 + a)x_t - a/bx_t^2 \tag{8.32}$$

and by redefinition of the constants

$$x_{t+1} = x_t(a' - b'x_t) \tag{8.33}$$

where

$$a' = (1 + a), \qquad b' = a/b \tag{8.34}$$

and then we drop the primes giving:

$$x_{t+1} = x_t(a - bx_t). \tag{8.35}$$

Here, we present May's main results without proof and concentrate on the way in which these can be applied. (See May, 1976, or Wilson, 1981, for the details.) There is an unstable equilibrium point $x^* = 0$, as in the continuous case. The equilibrium at

$$x^* = (a - 1)/b \tag{8.36}$$

is shown by May to be stable when

$$1 < a < 3. \tag{8.37}$$

He also shows that the condition

$$1 < a < 4 \tag{8.38}$$

must hold for 'sensible' behaviour. ($x_t \rightarrow 0$ if $a < 1$, and x_t diverges to $-\infty$ if $a > 4$.) We are left, therefore, with the question of what happens where a is between 3 and 4. The answer is as follows: for the lower part of the range, there is a single period oscillating solution between two points; then follows an oscillation between four stable points; and then, successively, oscillations between 2^n points for increasing n. This takes us to $a = 3.57$. For a between 3.57 and 3.85, there are more complicated oscillation involving sets of 3-cycles. And then for a between 3.85 and 4, the system characterized by x_t still oscillates but with no recognizable periodic structure, and this is called *chaotic* behaviour.

If we interpret these results in bifurcation terms, it means that there are successive values of the parameter a which are critical values at which the nature of the solution to the difference equation changes. $a = 3$ is an obvious one: there is a transition from a single stable equilibrium point to a periodic solution; and then between different kinds of periodic solutions; and, finally to divergence at $a = 4$.

What is the significance of these results? First, complex behaviour and bifurcation phenomena can arise from very simple looking equations — and this makes it all the more likely that this is true of more complex equations — as for example in Forrester's models, illustrated in the previous subsection. Secondly, the mathematics imposes conditions on one of the parameters, a in this case, in relation to modes of system behaviour. So, consequently and thirdly, it is interesting to interpret the meaning of such parameters in particular applications and to use the mathematical results in the specific cases.

May was mainly interested in ecological modelling. Difference equations are appropriate when the associated events are discrete, or where there are lags in responses which create a similar effect. In ecology, therefore, these methods have been applied to insect populations in cases where generations do not overlap: a cohort lives, dies, but leaves eggs which generate the next cohort. a is, in effect, a factor in the *increment* in egg-laying volumes from one generation to the next. If it is small enough to generate a population which is sufficiently less than the capacity b, then the equilibrium $(a - 1)/b$ will be stable. But if it is larger, different kinds of oscillations can appear, and these have been observed. Or if it exceeds 4, then extinction ('divergence to $-\infty$') will be the fate of the population.

Let us now consider shopping centre dynamics. The growth (or decline) of a centre can be taken as proportional (by a parameter ε) to the difference of revenue (D_j) and supply costs (kW_j) where k is taken as the unit cost in suitable time units.

$$\dot{W}_j = \varepsilon(D_j - kW_j). \tag{8.39}$$

We can replace $\dot{W}_j$ by

$$\dot{W}_j \simeq (W_{jt+\delta t} - W_{jt})/\delta t \tag{8.40}$$

and then, without loss of generality as before, take $\delta t = 1$, so

$$\dot{W}_j \simeq W_{jt+1} - W_{jt}. \tag{8.41}$$

(8.39) in difference equations form would then be

$$W_{jt+1} - W_{jt} = \varepsilon(D_j - kW_{jt}). \tag{8.42}$$

This is beginning to look like (8.33) (with W_{jt} for x_{jt}) except that there are no quadratic terms in W_{jt}. However, it turns out that an analysis similar to May's can be presented for (8.40), though without the periodic solutions occurring (Wilson,

1981). We will not pursue this here, however, but will concentrate on the analogy more directly by adding a W_{jt} factor to the right-hand side of (8.42):

$$W_{jt+1} - W_{jt} = \varepsilon W_{jt}(D_j - kW_{jt}). \tag{8.43}$$

The effect of this change is relatively minor: it amends the form of growth near the origin, but retains the same equilibrium points. Indeed some authors (Allen *et al.*, 1978, for example) use equations of the form (8.43) in preference to (8.42). We can then take W_{jt} from the left-hand side of (8.43) to the right to give

$$W_{jt+1} = (1 + \varepsilon D_j)W_{jt} - \varepsilon kW_{jt}^2. \tag{8.44}$$

This is now in the form of (8.33) with

$$a = 1 + \varepsilon D_j \tag{8.45}$$

and we can apply May's results directly. He showed that $1 < a < 3$ was the condition for stable equilibrium, and this implies

$$0 < \varepsilon D_j < 2. \tag{8.46}$$

Then, $3 < a < 4$ implied various periodic solutions and the corresponding condition is now

$$2 < \varepsilon D_j < 3. \tag{8.47}$$

We can now note two additional complications for this system, and then proceed to interpret these results. The first complication is that the revenue, D_j, is not, of course, a constant. It grows as W_j grows. However, the essence of May's analysis remains the same, but critical parameter values can be achieved and passed as W_{jt} (and hence D_j) grows. This has been confirmed by numerical experimentation (Beaumont, Clarke, and Wilson, 1980). The second complication is that D_j is also a function of all the other W_k's, $k \neq j$. A change in another centre will therefore affect D_j and will create the possibility of εD_j, for this reason, pursuing through a critical point. In systems analysis terms, this is a representation of feedback between subsystems, in this case shopping centres. We will not pursue this particular complication any further here.

The interpretation is then as follows: ε, as can best be seen from equation (8.40), determines the magnitude of response of entrepreneurs (or other corresponding decision makers) to excess profits, or to corresponding losses. That is, if $D_j - kW_j > 0$, $W_{jt+1} > W_{jt}$, and the increment, $W_{jt+1} - W_{jt}$ is proportionate to this and ε. Thus, the smaller ε, the more conservative the system is; the larger the more volatile. The mathematics bears out our intuition. In this case, therefore, the theory has offered some interesting ideas which can be tested in practice: can transitions from stable equilibrium to periodic solutions be identified in particular cases, for example? Are there cases where a system which was stable becomes unstable (i.e. when εD_j exceeds 3)? All this offers interesting avenues for further research.

One additional, more technical, point can usefully be made. In the above interpretation, we have assumed that there is a lag in entrepreneural response which makes the difference equation formulation appropriate, and this may indeed be a reasonable hypothesis. However, even if the differential equation was thought to be correct, it is often approximated in practice by a difference equation, and since parameters like a are implicitly proportional to the step length in such an approximation (through the assumption that $\delta t = 1$), then choosing a step length which is too large will be equivalent to a parameter which exceeds 3 or 4 and oscillations may appear which are *not* appropriate to the problem. The analysis here offers understanding of this and guidance on how small step length needs to be in particular cases.

8.4 DIFFERENTIAL EQUATIONS

8.4.1 Prey–predator systems

First, let us return to equations (8.12) and (8.13) of Section 8.1.4. They are repeated here for convenience:

$$\dot{x}_1 = (a - bx_2)x_1 \tag{8.48}$$

$$\dot{x}_2 = (cx_1 - d)x_2. \tag{8.49}$$

These are said to represent a prey–predator system since the rate of growth of x_1, say an animal population, diminishes with the growth of x_2, while the rate of growth of x_2 increases with the population of x_1. Thus, animal type-1 is the prey, and type-2 is the predator. This is a famous pair of equations whose discovery, made independently by two people, dates back to early in this century; they are known as the Lotka–Volterra equations.

It would take us beyond the scope of this book to show how the equations are solved, but it is possible to obtain some useful insights by examining characteristics of the solution graphically. Solutions to equations such as these can be presented as trajectories in a phase space diagram, as in Figure 8.7 for this particular case. The possible trajectories are ellipses, and one is shown as an example. This solution represents oscillations. At point A, for example, x_2 is at a minimum and so equation (8.48) shows that $\dot{x}_1$ is likely to be positive and x_1

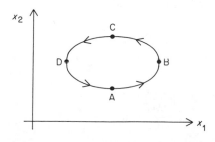

Figure 8.7 Ecosystem oscillations in phase space

increasing; this means, from (8.49), that x_2 will also be increasing. So we move to B. Here x_1 is at a maximum, so x_2 is still increasing, but beyond this x_1 starts to decrease; this gets us to C. Here x_2 is at a maximum, but begins to decrease and x_1 continues to decrease; and so to D. At this point x_1 begins to increase again while x_2 continues to decrease until the end of the cycle is reached with the return to A.

These equations have obvious applications within ecosystem analysis with x_1 and x_2 as animal populations at different trophic levels. The equations can be elaborated in various ways, and expanded to deal with a greater number of interactions. A general formulation might be

$$\dot{x}_1 = (a_{0i} + \sum_j a_{ij} x_j) x_i, \quad i = 1, 2, \ldots, N \tag{8.50}$$

for N species. a_{0i} is the natural rate of wrowth of species i and a_{ij} measures the interaction between species j and species i. This matrix therefore is nonzero in the places where 1's occur in the food-web matrix of Chapter 5, but is now equal to a quantity which measures the strength of the interaction. As usual, when $N > 2$, phase diagrams cannot be plotted and it is more difficult to understand the general behaviour of such systems.

8.4.2 Competition for resources

Consider now a more general version of equations (8.48) and (8.49) which may be written

$$\dot{x}_1 = M_1(x_1, x_2) x_1 \tag{8.51}$$

$$\dot{x}_2 = M_2(x_1, x_2) x_2. \tag{8.52}$$

M_1 and M_2 are each functions of x_1 and x_2 and are the growth rates associated with populations x_1 and x_2. If we took

$$M_1(x_1, x_2) = a - bx_2 \tag{8.53}$$

$$M_2(x_1, x_2) = cx_1 - d \tag{8.54}$$

then we would have the Lotka–Volterra equations. However, we wish to illustrate a different kind of argument: to see what can be deduced about the solutions of the equations from a consideration of very general properties of the functions M_1 and M_2.

Suppose the following conditions hold:

(i) If x_1 increases, then x_2 decreases; and vice versa. This can be expressed mathematically as

$$\frac{\partial M_1}{\partial x_2} < 0; \quad \frac{\partial M_2}{\partial x_1} < 0. \tag{8.55}$$

(ii) Neither x_1 nor x_2 can expand indefinitely. That is, there exists a constant K

such that

$$M_i \leqslant 0 \quad \text{if} \quad x_i > K. \tag{8.56}$$

(iii) If $x_1 = 0$, then x_2 grows to a certain point, say a_2, and then stops growing; and vice versa (with growth to a_1 say). That is

$$M_1(x_1, 0) > 0 \quad \text{if} \quad x_1 < a_1 \tag{8.57}$$

$$M_2(0, x_2) > 0 \quad \text{if} \quad x_2 < a_2. \tag{8.58}$$

The first point to note is that the Lotka–Volterra versions of M_1 and M_2 do not satisfy these conditions. Condition (i) is not satisfied, for example, since with M_2 from (8.54), $\partial M_2/\partial x_1 > 0$. A little thought shows that these three conditions are appropriate to two populations each competing for a fixed amount of some resource—plants for sunlight in our moorland ecosystem, animals for a food supply, sand so on.

On the basis of these three conditions, the trajectories in phase space of solutions to the equations can be sketched as in Figure 8.8. In this figure, trajectories are shown as dashed lines.

The continuous lines are plots of the curves

$$M_1(x_1, x_2) = 0 \tag{8.59}$$

$$M_2(x_1, x_2) = 0 \tag{8.60}$$

(sketched to show, *typically* what they would be like— they should only be really plotted when the functions were specified). Intersections of these two curves are possible 'steady states' (since $\dot{x}_1 = \dot{x}_2 = 0$ and the populations are fixed). It can be shown that B and P are the only two possible steady stable states (or if B is not

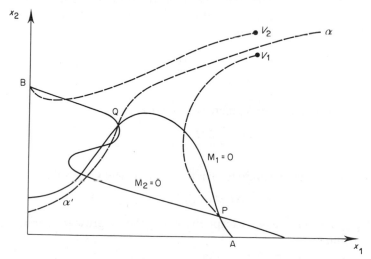

Figure 8.8   The equilibrium points in phase space of an ecosystem

stable, then A is—there are always two, one of them on an axis). Q is a 'saddle point'—a special kind of unstable equilibrium point, and plays a special role in that the two trajectories through Q, plotted as α and α', act as a *bifurcation curve*. Trajectories at points below $\alpha\alpha'$ tend to P; those starting above to B—as shown for two trajectories starting at V_1 and V_2 respectively.

At least two interesting insights have emerged from this analysis which can now be emphasized. First, there are only two stable states. One of these, P, implies coexistence of the two populations since both x_1 and x_2 are nonzero; the other, B, implies that population -1 is wiped out in the competition (because x_1 is zero at B; or vice versa if A is the stable point). This paucity of stable points, and the nature of them, is interesting for such a potentially complicated problem. Secondly, we can examine the effect of the bifurcation curve: if the system is disturbed from one of its stable states, say from P to near V_1, then it is in a critical region. At V_1 it will return to P; but if the disturbance takes it to V_2 across the bifurcation line, or an additional small impulse does so, then it will go to B and one of the populations will be annihilated.

As already noted, this example has obvious applications in ecosystems analysis to both animal and plant populations. It also turns out to have interesting applications in urban systems. For example, x_1 and x_2 may represent the amounts of floorspace in two different kinds of shops (say corner shops and supermarkets respectively). These shop types compete for a fixed resource for them—the availability of shoppers' expenditure. Point P represents coexistence; point B represents the annihilation of corner shops. If a planner considers his task is to preserve both kinds of shops, he needs to know where the critical region in the neighbourhood of $\alpha\alpha'$ is located and to ensure that it is not crossed. For a second urban example, x_1 and x_2 may be taken as the shares of two transport modes competing for a fixed supply of passengers. A similar analysis then applies.

As with the Lotka–Volterra equations, the two-population equations (8.51) and (8.52) can be generalized to N-populations. The equations become

$$\dot{x}_1 = M_i(x_1, x_2, \ldots, x_N)x_i, \quad i = 1, 2, \ldots, N \tag{8.61}$$

and the three conditions become:

(i) x_j decreases if any other x_k increases. That is,

$$\frac{\partial M_j}{\partial x_k} < 0. \tag{8.62}$$

(ii) x_i cannot expand indefinitely. That is, there exists k such that

$$M_i \leqslant 0 \quad \text{if} \quad x_i > k. \tag{8.63}$$

(iii) If all populations are zero except one, say x_i, then x_i grows to a limit and stays at that limit:

$$M_i(0, 0, \ldots, x_i, 0, \ldots, 0) \begin{array}{l} > 0 \text{ if } x_i < a_i \\ < 0 \text{ if } x_i > a_i. \end{array} \tag{8.64}$$

As usual, the problem is that phase space diagrams can no longer be plotted for $N > 2$, though we can be sure that only a small number of stable points exist and that there are critical regions near bifurcation curves. This is particularly interesting in relation to the N-region shopping model. Consider the differential equation for shops in j, defined earlier in equation (8.39).
If we write D_j explicitly as $\sum_i S_{ij}$, this is:

$$\dot{W}_j = \varepsilon \left[\sum_i S_{ij} - W_j p_j \right] \tag{8.65}$$

say. If we define

$$M_j(W_1, W_2, \ldots, W_N) = \frac{\varepsilon \left[\sum_i S_{ij} - p_j W_j \right]}{W_j} \tag{8.66}$$

then we see that (8.65) can be written as

$$\dot{W}_j = M_j(W_1, W_2, \ldots, W_N) W_j. \tag{8.67}$$

Further, if we substitute for S_{ij} in (8.66), recalling that

$$S_{ij} = \frac{e_i P_i W_j^\alpha e^{-\beta c_{ij}}}{\sum_j W_j^\alpha e^{-\beta c_{ij}}} \tag{8.68}$$

then after some rather laborious calculus and algebra, it can be shown that the M_j in (8.66) satisfy the conditions (i)–(iii) in equations (8.62)–(8.64) above. This is perhaps not surprising, as the N shopping centres are obviously competing for a fixed resource—the shoppers' expenditure. However, it is nice to know that the usual model does satisfy these conditions and to be alerted to the possibility of the existence of interesting dynamic behaviour for a system governed by equations (8.65). That is, there will be a small number of possible stable points and a disturbance may take the system into the neighbourhood of critical regions.

8.5 A MICRO-SIMULATION APPROACH TO DYNAMICS

8.5.1 Static preliminaries

In most of the models used in this book, we have defined variables like T_{ij} or x^{mnk} to describe the system of interest. More generally, suppose we define $T(x_1, x_2, x_3, \ldots, x_N)$ as the number of system elements with characteristics $(x_1, x_2, x_3, \ldots, x_N)$. If the different characteristics have $n_1, n_2, \ldots, n_N$ cells, or categories, then the array $T(x_1, \ldots, x_N)$ has $\prod_{k=1}^{N} n_k$ possible cells, and typically this is a very large number, especially if N is large.
An alternative representation is to list the system elements and their

characteristics. If we add a superscript, r, to x_i to label system element r, then the system is described by $(x_1^r, x_2^r, x_3^r, \ldots, x_N^r)$, $r = 1, \ldots, M$, say, if there are M elements in total. There are μN such numbers in the system description and although this also is usually large, it is often much smaller than $\prod_{k=1}^{N} n_k$. (This is because the T-array usually has a large number of zero entries in it.) This explains why the arrays we use usually have relatively few subscripts and superscripts: $\prod_{k=1}^{N} n_k$ becomes too large too soon. The alternative representation, described above, has a history going back to the 1950's but has only been used relatively infrequently. The simplest way to use it is in conjunction with Monte Carlo simulation methods — and hence the name micro-simulation model. Each characteristic, for each element r, x_i^r, is estimated in relation to a known conditional probability distribution using random numbers in such a way as to ensure that the probability distributions are satisfied overall.

Formally, we can write (following Wilson and Pownall, 1976):

$$x_i^r = x_i^r \left[P_i(x | \ldots), R_i^r, \Gamma \right] \tag{8.69}$$

where P_i is the probability of x_i^r taking the value x conditional on whatever follows the bar. R_i^r is the random number and Γ is a symbol representing any further constraints which may restrict the assignment — like overall budget constraints. The model has a causal structure which is represented by the pattern of conditional probability distributions used. It is partly reflected in the order in which they are computed, since any x_j^r which appear on the right-hand side of (8.69) must be computed before x_i^r.

The kind of model described is essentially static and we have introduced it here to make the representation available. In the next subsection, we briefly outline its role in dynamical analysis.

8.5.2 Dynamics and micro-simulation

We have seen at various earlier points in the book that accounting notions lead us to consider differential or difference equations of the following forms:

$$\frac{dN_i}{dt} = \sum_k (a_{ki} N_k - a_{ik} N_i) \tag{8.70}$$

or

$$N_{it+1} - N_{it} = \sum_k (a_{ki} N_{kt} - a_{ik} N_{it}). \tag{8.71}$$

Here, the array $\{a_{ik}\}$ consists of transition coefficients: a_{ik} is the probability of an element shifting from state i to state k in a unit time period. But now suppose each i and k, which describe system elements, are subscript lists. That is, they could be the $(x_1, x_2, \ldots, x_N)$ of the previous subsection. We thus have the same problems as before with the state variable N, but they are compounded for the transition coefficients which now become $a(x_1, x_2, \ldots, x_N; x'_1, x'_2, \ldots, x'_N)$ and are typically of very high dimensionality indeed.

Micro-simulation helps to solve this problem as in the static case, but the probability distributions involved now relate to transitions in time. Usually, the problem will be broken down by having a component probability relating to whether a characteristic will change or not, and then if it will, the probability of what it will change to. The calculation then proceeds using Monte Carlo methods as before and produces computational solutions to dynamical problems which may be otherwise intractable.

8.6 CATASTROPHE THEORY

8.6.1 Introduction

Catastrophe theory is concerned with certain aspects of systems which are described by endogenous state variables, $\mathbf{x}$, and exogenous variables, $\mathbf{u}$, and for which there exists a function, $E(\mathbf{x}, \mathbf{u})$ say, such that the system state $\mathbf{x}$ is that which minimizes E, given $\mathbf{u}$. Let $\hat{\mathbf{x}}(\mathbf{u})$ denote the value of $\mathbf{x}$ for which E is a minimum for given $\mathbf{u}$. Then, as $\mathbf{u}$ varies (that is, as the environment of the system changes), $(\hat{\mathbf{x}}(\mathbf{u}), \mathbf{u})$ traces out a *surface* in $(\mathbf{x}, \mathbf{u})$ space. If there is one $\hat{\mathbf{x}}$, $\hat{x}_1$, and two $\mathbf{u}$, say u_1 and u_2, then $(\hat{x}_1, u_1, u_2)$ traces out a surface in three-dimensional space. For higher dimensions, it is impossible to visualize geometrically, of course. These surfaces (or curves in the 2-D case) are usually single sheets with a unique $\hat{\mathbf{x}}$ for each given $\mathbf{u}$, as shown for two dimensions in Figure 8.9(a); it is possible, however, that the surface may be folded, as in Figure 8.9(b): between $u_1 = u'_1$ and $u = u''_1$, there are three possible values of x_1 corresponding to a single u_1. Catastrophe theory is particularly concerned with such cases.

When the exogenous variable u_1 changes smoothly, we usually expect correspondingly smooth changes in $\hat{x}_1$. However, in the cases where the surfaces are folded, there may be discrete jumps as $\mathbf{u}$ changes smoothly near some critical region. Catastrophe theory was invented by Thom and he showed that there are only seven distinct types of elementary catastrophe, each one characterized by the number of exogenous variables and the form of the function, E. Most of the

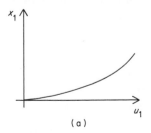
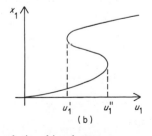

Figure 8.9 (a) A one–one relationship between a variable and a parameter; (b) a multi-valued relationship

Table 8.1 The seven elementary catastrophes

Name	End. variables	Control variables	Potential function
Fold	1	1	$\frac{1}{3}x_1^3 + u_1 x_1$
Cusp	1	2	$\frac{1}{4}x_1^4 + \frac{1}{2}u_1 x_1^2 + u_2 x_1$
Swallow-tail	1	3	$\frac{1}{5}x_1^5 + \frac{1}{3}u_1 x_1^3 + \frac{1}{2}u_2 x_1^2 + u_3 x_1$
Hyperbolic-umbilic	2	3	$\frac{1}{3}x_1^3 + \frac{1}{3}x_2^3 + u_1 x_1 x_2 - u_2 x_1 - u_3 x_2$
Elliptic-umbilic	2	3	$\frac{1}{3}x_1^3 - \frac{1}{2}x_1 x_2^2 + \frac{1}{2}u_1(x_1^2 + x_2^2) - u_2 x_1 - u_3 x_2$
Butterfly	1	4	$\frac{1}{6}x_1^6 + \frac{1}{4}u_1 x_1^4 + \frac{1}{3}u_2 x_1^3 + \frac{1}{2}u_3 x_1^2 + u_4 x_1$
Parabolic-umbilic	2	4	$\frac{1}{2}x_1^2 x_2 + \frac{1}{4}x_2^4 + \frac{1}{2}u_1 x_1^2 + \frac{1}{2}u_2 x_2^2 - u_3 x_1 - u_1 x_2$

results are independent of the form of the function, however. It is customary to take polynomial functions to illustrate the mathematical analysis, as we do below, but the form of the catastrophes remains the same for a wide class of functions in some local sense, determined only by the number of control variables.

The seven elementary catastrophes are named and have the potential functions shown in Table 8.1.

8.6.2 The fold catastrophe

The fold is the first and simplest of Thom's catastrophes with a single state variable and a single control variable. The potential function is

$$E = \frac{1}{3}x_1^3 + u_1 x_1. \tag{8.72}$$

x_1 is the state variable—an endogenous variable—and u_1 is the control variable—an exogenous variable. The system takes the value of x_1 which arises when E is minimized. E is a maximum or a minimum (i.e. takes a 'stationary' value) when

$$\frac{dE}{dx} = x_1^2 + u_1 = 0. \tag{8.73}$$

Thus,

$$x_1 = \pm\sqrt{-u_1} \tag{8.74}$$

at such points.

$$\frac{d^2 E}{dx^2} = 2x_1 \tag{8.75}$$

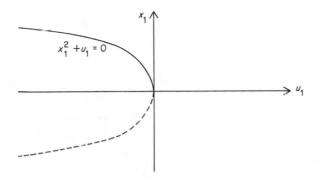

Figure 8.10 The fold catastrophe

and since this is positive for x_1 positive, the positive root of (8.74) is the minimum value of E; and, conversely, the negative root is a maximum. Note from (8.74), however, that the roots only exist when $u_1 \leqslant 0$.

This information is plotted on Figure 8.10—essentially by noting that (8.73) is the equation of a parabola and that the top half, the solid curve, represents the minima and hence stable points, while the bottom half, the dashed curve, represents maxima and hence unstable points.

If u_1 changes smoothly from some negative value, then x_1 will decline. As $u_1 \rightarrow 0$, x_1 will decline with increasing rapidity. As u_1 passes through zero to a positive value, the system as described by (8.72) has no state available to it and so 'disappears'. In practice, as we will see in an example below, this often means a jump to a state described by other equations.

8.6.3 The cusp catastrophe

Thom's second kind of catastrophe, the cusp catastrophe, involves the minimization of

$$E = \tfrac{1}{4}x_1^4 + \tfrac{1}{2}u_1 u_1^2 + u_2 x_1 \tag{8.76}$$

for a state variable x_1 and exogenous variables u_1 and u_2. The stationary values (maxima or minima) of E are found by solving

$$\frac{\mathrm{d}E}{\mathrm{d}x} = x_1^3 + u_1 x + u_2 = 0. \tag{8.77}$$

A cubic equation can have either three real roots or one. It can be shown that the condition for three real roots is

$$(-\tfrac{1}{3}u_1)^3 > (\tfrac{1}{2}u_2)^2 \tag{8.78}$$

and this, of course, implies

$$u_1 < 0. \tag{8.79}$$

The boundary of the region defined by (8.78) is

$$4u_1^3 + 27u_2^2 = 0 \tag{8.80}$$

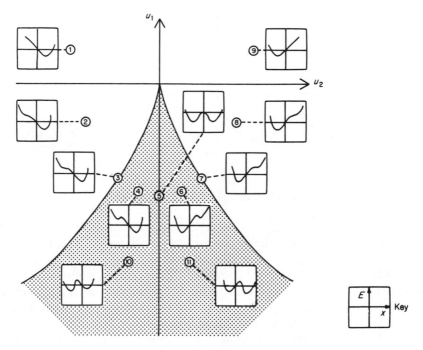

Figure 8.11 Potential functions for different combinations of parameter values: the cusp catastrophe

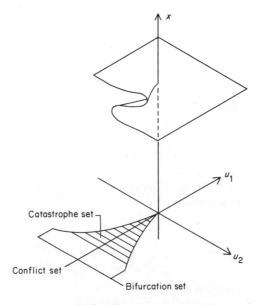

Figure 8.12 The cusp catastrophe

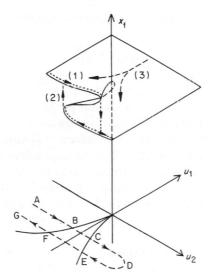

Figure 8.13 The cusp catastrophe with sample trajectories

and this is the cusp-shaped curve plotted in the (u_1, u_2)-plane in Figure 8.11. The shaded region inside the cusp is the region of (u_1, u_2) values which satisfy (8.78). At various points in this plane, using what appears to be a standard method of the catastrophe theory mathematicians, we show what the plot of E against x looks like for those u_1, u_2 values. Inside the cusp region, it is easy to see that there are two minima and one maximum; outside the region, one minimum only. In order to plot the surface (x_1, u_1, u_2) values for which E is a minimum, we now move to three dimensions. It is easy to see that to the left of, and outside, the cusp region, $\hat{x}_1$ is positive, and negative for right and outside. Inside the region, there are two minima so that the surface must be folded. This is sketched in Figure 8.12. The cusp region in the (u_1, u_2) plane can now be seen as the projection of the folded part of the surface onto the (u_1, u_2) plane.

As with the fold catastrophe, we can see how this can play a role in dynamical systems theory. u_1 and u_2 are external to the system and exogenous change in these variables is a path in the (u_1, u_2) plane which traces a path on the surface. The new phenomena occur near the edges of the cusp region which turns out to be a critical region. Consider the path AD in the (u_1, u_2) plane in Figure 8.13 and the corresponding curve on the surface in the figure. At B, the cusp region is entered, but the system stays in an 'upper surface' state. At C, as it crosses the cusp region at the other side, however, as u_2 continues to increase, it must jump to the lower surface. The path can then be retraced straightforwardly, but particularly noting that the jump back takes place at F—again the far side of the cusp region. There is an 'hysteresis effect'. It can also be seen, by comparing back to the sample plots on Figure 8.11, that the minimum occupied by the system literally disappears at a jump point. In terms of delay convention, we have assumed that perfect delay operates in this case.

8.6.4 Facility size versus distance travelled: an example using the fold catastrophe

This example can be seen as a very general problem in a number of fields. We have already seen a version of it in Section 8.2.1 involving the spatial distribution of shopping centre sizes. For convenience, we consider the same problem again, but bearing in mind that it is an archetype of many others. The essence of the general problem is: suppose the per capita benefits to be gained, on average, increase with the size of the facility to be provided, but decrease with the distance, on average, which has to be travelled to reach it. The extreme solutions are first, one large facility with people travelling the largest maximum distance to reach it, and, secondly, many small facilities with a short associated average distance of travel. How do we strike a balance between the benefits of size and the costs of travel? How do we choose a solution between the two extremes (or indeed at one of them)?

We examine the problem at a rather coarse level of resolution, not, for once, distinguishing spatial detail. Let W be the average size of a shopping centre (or other facility) and c the average cost of travel. As indicated in looking at the extreme solutions, we expect a functional relationship between W and c: the greater c, the greater W. For the time being, therefore, we will assume they are monotonically related and we will work with c as the endogenous state variable of the system. (For example, if average centre size is small, average distance travelled will also be small, and vice versa.)

The next step in the argument is to find a mechanism for determining the size of c. We propose that this is determined by utility maximization, where utility is made up of the two components discussed earlier: u_1 is the average benefit from facility size, u_2 from travel. We expect u_1 to increase with W, and hence c, and u_2 to decrease, as indicated in Figure 8.14. We assume that u_1 takes a logistic form and increases to some maximum, and the u_2 is backward-sloping linear and, of course, negative. Total utility is

$$u = u_1 + u_2. \tag{8.81}$$

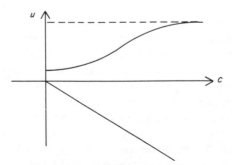

Figure 8.14 Benefit and cost for a shopping centre problem

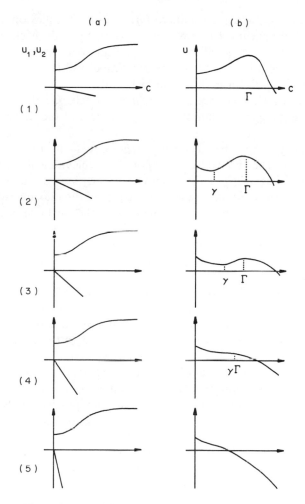

Figure 8.15 Utility functions with changing 'ease of travel'

We characterize the main exogenous variable for this system by the gradient of the travel line, say $-\beta$, and take this as a measure of the 'ease of travel' at a particular time. So the next task is to investigate the changing shape of u as β changes. In Figure 8.15, we show (a) u_1 and u_2 separately with β steadily increasing, and (b) $u = u_1 + u_2$ for each case. Low β means that travel is relatively easy, and so the sequence from top to bottom shows travel becoming increasingly difficult; from a historical point of view, therefore, the figure should be read from bottom to top. We can now examine column (b) and the behaviour of the maximum of u. In case 1, there is a unique maximum at Γ. In case 2, a *local* maximum has appeared at $c = 0$, a local minimum at $c = \gamma$, but $c = \Gamma$ remains the global maximum. By case 3, $c = 0$ has become the global maximum and Γ the local one. At case 4, γ and Γ coincide to form a point of inflexion and the local maximum at Γ disappears. The continuation of this process is clearly visible in case. 5. Thus high β means difficult travel and therefore many small shopping centres, the $c = 0$ solution; low β means easy travel and large shopping centres corresponding to $c = \Gamma$. Cases (2)–(4) show that there is a range of β values, say $\beta_0 \leqslant \beta \leqslant \beta_1$ for which there are two maxima. Which one the system adopts depends on the direction of change of β over time and the delay conventions adopted.

This information can be succinctly represented on one figure—as in Figure 8.16 below. The vertical axis is a plot of $\hat{c}$—the value of c at which u is a maximum, plotted against β. This can now be clearly seen as an example of the fold catastrophe. The continuous curve is the plot of the maxima (which are the stable points in this example); the dashed curve the unstable minimum. The dotted and dashed-and-dotted lines are possible system trajectories for increasing β and decreasing β respectively. Perfect delay convention has been assumed, so the jump takes place at the last possible moment, though this can easily be modified if necessary.

This can be related to the formal mathematics of the fold catastrophe presented in Section 8.6.2, and it illustrates an important point which was just touched on previously. The folded curves in Figures 8.10 and 8.16 are, with a shift of origin of coordinates, obviously of the same topological shape. In this example, however, we have added a new set of states: the $c = 0$ stable maxima. This creates a region, as we saw, where there are two possible stable states, but perhaps more importantly, provides a state for the system to jump into when it must leave the upper curve (which is equivalent to u becoming positive in Section 8.6.2, Figure 8.10). This is a case, therefore, where *some* of the system states are described by the fold catastrophe, but others are added.

We can now investigate how this analysis connects to the conventional model of shopping trips we have used on a number of occasions earlier. This model can be taken as

$$S_{ij} = A_i e_i P_i W_j^\alpha e^{-\beta c_{ij}} \tag{8.82}$$

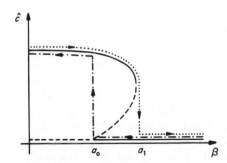

Figure 8.16 The fold catastrophe for the shopping-centre-size example

where

$$A_i = \frac{1}{\sum_j W_j^{\alpha} e^{-\beta c_{ij}}}$$

(8.83)

with the usual definition of variables (see, for example, p. 65). Since

$$W_j^{\alpha} = e^{\log W_j^{\alpha}} = e^{\alpha \log W_j}$$

(8.84)

the last two terms can be combined as

$$e^{\alpha \log W_j - \beta c_{ij}} = e^{\mu b_{ij}}$$

(8.85)

say, where we have defined a term b_{ij} which can be taken as a measure of benefit. Without loss of generality, we can take μ to be 1 (since otherwise it is a scaling factor which applies throughout) and take

$$b_{ij} = \alpha \log W_j - \beta c_{ij}.$$

(8.86)

In effect, we have

$$u_{1ij} = \alpha \log W_j$$

(8.87)

as size benefits and

$$u_{2ij} = -\beta c_{ij}$$

(8.88)

as travel benefits. They can be plotted in the maner of Figure 8.14 and this is done in Figure 8.17. The subscripts have been dropped again for convenience. The difference is in u_1 of course. Because $\log W_j \to -\infty$ as $W_j \to 0$, the local maximum at $c = 0$ never develops. So in this case, there is always a unique maximum which moves smoothly as β change smoothly. Essentially, this is because the u_1 curve in Figure 8.17 has no point of inflexion like the logistic curve in Figure 8.14.

This analysis leads to a possible change in the conventional shopping model. If the real world *does* exhibit jump behaviour in shopping centre sizes (and this is a matter of empirical research), then since we have seen that W_j^{α} as an attractiveness factor does not produce such behaviour, there would be a case for replacing the power function by a logistic function: $\alpha \log W$ would be replaced by, say,

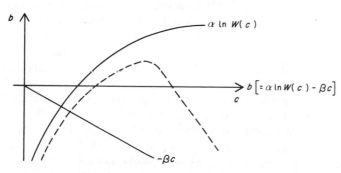

Figure 8.17 Components of utility in the conventional model

$\delta/[1 + \gamma \exp(-\varepsilon W)]$, where δ, γ, and ε are constants to be determined. The model equation (8.82) would then become

$$S_{ij} = A_i e_i P_i \exp \left\{ \frac{\delta}{[1 + \gamma \exp(-\varepsilon W_j)]} \right\} \exp(-\beta c_{ij}) \qquad (8.89)$$

with a corresponding adjustment for A_i:

$$A_i = \frac{1}{\sum_j \exp\{\delta/[1 + \gamma \exp(-\varepsilon W_j)]\} \exp(-\beta c_{ij})}. \qquad (8.90)$$

For this model, we can conjecture that smooth changes in β would lead to jumps in some W_j's if this model was embedded in an optimizing procedure such as that of Section 6.3. However, this problem is very difficult to handle analytically as we now have N state variables instead of the single (albeit possibly 'typical' one) of the fold catastrophe analysis above. Numerical experiments are possible, however.

8.6.5 The cusp catastrophe and modal choice

We now turn to an example governed by two exogenous variables, u_1 and u_2. There is still a single state variable, x_1, which now denotes choice of travel mode by an individual. Let $x_1 < 0$ denote the choice of mode 1 and $x_1 > 0$ that of mode 2. An interesting model can then be constructed based on the mathematics of the cusp catastrophe described in Section 8.5.3. We take u_1 to be a 'habit' factor and it will turn out that $u_1 > 0$ will designate 'no habit formed', and increasingly greater negative u_1 will designate increasing degrees of habit. u_2 is taken as proportional to the cost differences between the two modes:

$$u_2 = k\Delta c = k(c_2 - c_1). \qquad (8.91)$$

We can now interpret Figure 8.13 for these definitions of u_1 and u_2. For a given individual, assume that u_1 has a fixed value. This determines a plane perpendicular to the u_1 axis which cuts the x_1-u_1-u_2 surface in a curve as depicted in Figure 8.18. Such a cross-section can be considered as a plot of x_1 against u_2, as shown, for given values of u_1. Case (a) shows a plot for negative u_1, when the plane intersects a folded part of the curve; case (b) for positive u_1, 'above' the fold. The dashed lines show possible trajectories, with the perfect delay convention assumed to operate, as Δc changes. Consider case (1). When Δc is large and negative, $c_2 \gg c_1$ and so mode 1 is the only choice. As Δc is reduced, there is a region $-\gamma < \Delta c < \gamma$ where either mode is seen to be possible, but with the perfect delay convention operating there is no jump until $\Delta c = \gamma$. Case (2) shows the symmetrically opposite case. γ is a parameter which determines the 'width' of the hysteresis effect; it is determined by the value of u_1, of course, the 'habit' factor. It is now clear from Figure 8.18(b) why $u_1 > 0$ implies no habit, or inertial factor: as soon as Δc changes sign, such a person changes mode.

This argument has been conducted at the micro level. To turn it into a useful

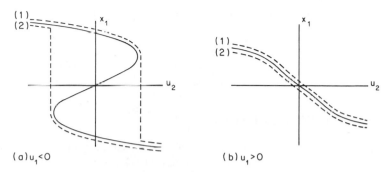

Figure 8.18 Modal choice, with and without hysteresis

aggregate model, the distribution of u_1 in the population would have to be assumed. It may then be possible to add hysteresis effects to existing modal split models. This would be a complicated operation because of the region of Figure 8.18(a) where a person could be in one of two states. An assumption, based on the *previous history of* Δc, would have to be made about the distribution of people between the two states.

8.7 BIFURCATION: CONCLUDING COMMENTS

This chapter is one of the most important in the book; all interesting systems change and evolve, and the methods of dynamical systems analysis are therefore crucial. Unfortunately, a detailed mathematical treatment would take us beyond the scope of the book and it has been possible to present a sketch only. The reader whose appetite for dynamical theory has grown can pursue the subject further through the literature cited in the 'Notes on further reading' below. In this section, however, we highlight a particular feature of all the methods presented above and note its importance for modelling and analysis in the future: bifurcation.

In comparative static analyses, we are concerned with equilibrium points and their stability, and we saw in Section 8.2 how parameters of models can have critical values at which the nature of the equilibrium can change — for example, to become unstable so that the system has to jump to a new state — a new form of shopping centre structure, or a new order in the network hierarchy for example. In Section 8.6, we saw how catastrophe theory suggests a broader framework for this kind of thing, offering the possibility of modelling more complex behaviour than we would intuitively think possible and, importantly, offering the basis of an explanatory model too.

But difference equations and differential equations represent more general model systems. Catastrophe theory is restricted to so-called 'gradient' systems. We saw in Sections 8.3 and 8.4 that there can be a variety of solutions to such equations: stable equilibrium, unstable equilibrium, periodic, and so on. In this context, bifurcation is concerned with critical parameter values at which the nature of the solution, *for a particular model or system*, changes. We presented dramatic examples of this in Section 8.3.2 for example, where stable equilibrium solutions became periodic in ecological or shopping centre contexts.

The general point to emphasize is that the study of *criticality* is important for both analysis and planning, and sometimes in relation to the evolution of systems. This is an emphasis to be borne in mind for much future research.

NOTES ON FURTHER READING

A general text on dynamical systems theory is *Hirsch and Smale (1974). Useful texts on catastrophe theory are by *Zeeman (1977) and *Poston and Stewart (1978) following the classical book of **Thom (1975). A wide range of applications in urban and regional analysis is contained in *Wilson (1981) and a useful collection of papers is *Martin, Thrift, and Bennett (1979). *Forrester's (1969) *Urban dynamics* provides a substantial example of sets of difference equations (8.3.1.) The simple example which leads to complicated behaviour (8.3.2) is based on the work of †May (1976). The account of prey–predator and competition-for-resources differential equations is based on, and can be supplemented by, †Maynard Smith (1974, especially Chapters 2 and 5). The original work in micro-simulation modelling was by *Orcutt *et al.* (1961); a more recent example is *Clarke, Keys, and Williams (1981) (see Contents page and section on simulation modelling in Section 8). A treatment of the more conventional methods of dynamical analysis, such as the calculus of variation, can be found in *Miller (1979).

CHAPTER 9

Control and planning

9.1 Introduction

Control and planning are closely related concepts which in a certain colloquial sense can mean the same thing. Here, however, we use 'control' in a rather technical sense. Consider variables associated with systems which can be controlled by another subsystem or by an outside agent. These would include, for example, the frequency of burning of a moorland stand of heather, the setting of a value which released water from a reservoir, or a local tax rate in a city. The subsystem or agent which fixes such values is part (or all) of a *control system*. Typically, control systems are subsystems within feedback loops. They receive information on the outputs of the system being controlled and then fix one or more of the input variables to the system to values which analysis has shown will lead to desired outputs. The concept has its clearest technical meaning in the engineering field and there are many familiar examples in a rapidly-developing microelectronic age. But the concept also has application in a wider context — as a component within a planning system for example. Thus, building in Britain is regulated by the 'development control' section of the local planning authority, acting in accord with objectives which are supplied to it. The whole planning system can be considered to be a control system for cities and regions, thereby illustrating another point that control systems can themselves have subsystems which also have control functions. Indeed, in general, a complicated system may have many control subsystems.

'Planning' is a much wider concept and covers the whole variety of its colloquial meanings: preparing means for problem solving, to achieve certain objectives, or in anticipation of future events.

In Section 9.2, we discuss various aspects of control, including some ideas and theorems which have quite wide applicability. In Section 9.3, we similarly discuss the typical components of planning processes. In Section 9.4, some of the formal tools of planning are outlined. Planning in practice can only be discussed in terms of the social and institutional frameworks within which it is practised and these are discussed in general terms in Section 9.5. Some particular examples of such

211

institutions are outlined in Section 9.6. Some concluding comments are offered in Section 9.7.

9.2 ASPECTS OF CONTROL

9.2.1 Introduction

In some texts on systems analysis (for example, Bennett and Chorley, 1978, where the methods, as in this book, are applied to the environmental sciences), the notion of control systems are made central to the whole argument. In others (as in the framework in Quade *et al.*, 1976), it is very much a subsidiary concept absorbed under the heading of 'optimization' as one of the tools of the systems analyst. When the first perspective is adopted, it is usually in the context of formal mathematical control theory which we mentioned in Chapter 6. Here, we take the view that our knowledge of environmental systems is both imperfect (and hence inadequate for the formal theory) and not in canonical form (thus making the methods of the formal theory difficult to apply in practice), but that it is a concept of some significance. This arises for at least two reasons. In our general concern with planning, we are inevitably interested in concepts like controllability but also in general notions of control system synthesis—the problem unique to systems theory mentioned in Chapter 2. We therefore need to be able to consider any general theorems about controllability on the one hand and to explain by whatever means possible ways of understanding what makes control systems as such useful or efficient; and to base control-system design principles on this understanding. In the next subsection, we illustrate the first of these ideas with a brief discussion of Ashby's law of requisite variety (Ashby, 1956). The second is more difficult because of the enormous variety of types of control system, but we focus on an unorthodox but interesting approach due to Beer (1972, 1980) based on a detailed analysis of the human nervous system as a control system. This provides insights and analogies for many other situations. Other aspects of control will then be dealt with under the various 'planning' topics which follow.

9.2.2 The law of requisite variety

We introduced the concept of variety in Chapter 2 as a measure of the complexity of a system. It is the number of possible states a system can get into. Ashby (1956) introduced his law of requisite variety in relation to control systems: that the control system must have at least as much variety as the system it is intended to control. This has an immediate intuitive appeal, given the definition of variety, and some important practical consequences for the design of control systems within planning. A small group of city planners, for example, are unlikely to have in their own resources enough variety to control a city of a million people. However, various strategies are open to them. First, they can recognize that some of the control functions can be taken on by other organizations within the city. This includes, for example, the many organizations which are concerned with running and developing the economy of the city. They can then try 'higher-order' control by regulating these lower-order controllers in some general way without having to concern themselves with the detail. In a market economy, for example, the operation of the market (expansion through profit, deletion by bankruptcy) provides the basic control and this can be regulated in some overall way—for example through interest rates. Secondly, they can try to increase their own systems variety by intellectual and technological means—for example by building computer models of the city which provide them with much more information. Thirdly, they can try to restrict the city's variety by imposing regulations which limit it—and this, of course, is a strategy which, in broader contexts, is forced on to dictatorial governments.

The notion of generating variety within and for an effective control system is an important element of Beer's ideas, to which we now turn.

9.2.3 The central nervous system as a 'model' control system

Beer's (1972) argument is essentially that the most sophisticated and successful control system to evolve in nature is that of the control nervous system, and therefore if we analyse it in depth our understanding of such success may help in the design of other kinds of control systems. He has recently conducted a similar argument from first principles, leading to similar broad conclusions (Beer, 1980).

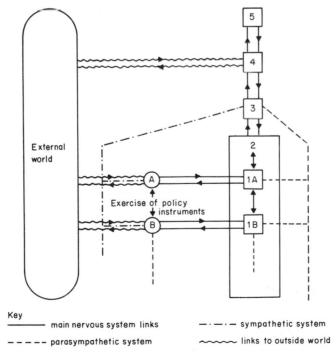

Figure 9.1 A CNS-type control system (Beer, 1972)

214

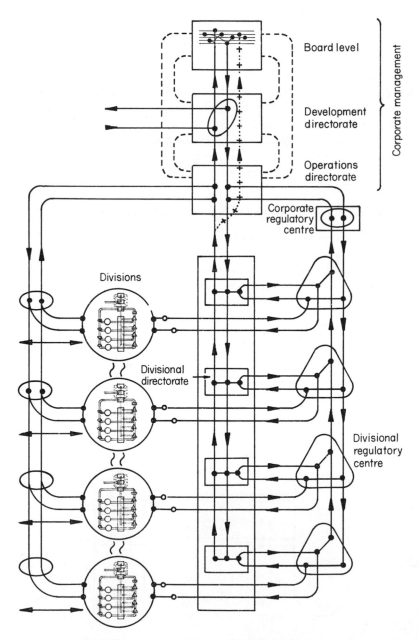

Figure 9.2 Corporate structure and a CNS-system

He argues that an organization (let us call it) with basic activities $A, B, C, \ldots$ is controlled by a five-level structure. This is shown diagrammatically in Figure 9.1. The organization is linked to the external world through the units which carry out the activities and also through level 4.

Broadly speaking, the control functions of the different levels can be described as follows: level 5 is the policy-making unit; level 4 is concerned with receiving information from outside and with transmitting it (including decisions on *what* to transmit) both up and down; level 3 is the overall management level for autonomous functioning (i.e. given a policy); level 2 coordinates the level 1 controllers; and at level 1, the particular activities are managed. There are two other kinds of links shown on the diagram: the sympathetic and parasympathetic nervous systems, which are concerned with emergency excitation and inhibition respectively.

In the case of the human body, the activities $A, B, C, \ldots$ are the basic actions; levels 1 and 2 are managers and coordinators; level 3 controls the whole autonomic nervous system—breathing rate, heart-beat, and so on. Levels 1 to 3 are mostly located on the spine. Level 4 is sited at the lower part of the brain. It receives the information from the senses and passes information (including decisions from level 5 and problems to level 5) up and down. Level 5 is the main, upper, part of the brain. It is easy to see how the control system of a firm can be modelled in the same way, and this is shown explicitly in Figure 9.2.

Beer's argument is essentially that the control system of any organization can be modelled in this way—even if, in some cases, particular parts of it are missing. Indeed, one of his main arguments is that the intelligence and switching functions performed by level 4 are often missing from many kinds of organization and that this can lead to many problems—mainly from lack of adequate information or information overload (lack of requisite variety in either case) at level 5.

It is interesting to note in passing that Miller (1978), who gives great emphasis to the notion of subsystems which are 'deciders' rather than 'controllers' and who does not mention requisite variety explicitly, nonetheless gives great attention to the consequences of information overload in systems.

These ideas will be explored more explicitly in Section 9.6, in the context of examples of planning systems.

9.3 ASPECTS OF PLANNING: SOME BASIC CONCEPTS

Planning is concerned with finding good ways of solving problems, or achieving goals or objectives. An immediate question is: whose problems, whose goals? In most particular situations, there will in practice be major conflict. We address some of the questions involved in the next few sections. However, it can be argued that for a wide range of value systems, there will be forms of planning which are appropriate and that the kinds of activities involved are very much the same in each case.

Consider three kinds of possible broad underlying value system. Someone believing in benefits of a free-market economy might wish to plan to deal only with acknowledged aspects to market failure. Alternatively, Miliband (1976) argues that the three main objectives of socialism can be summarized as concerned with equity, democracy, and efficiency. A third kind of broad objective is a concern with 'freedom'. One of the first features to arise from the application of systems analytical planning concepts to any of these worlds is that much more explicit notions of the overall objectives are required, and this alone can be beneficial activity.

In any of these situations, or others, what can be planner do? First, assume we are talking about public planning: the planner as an agent of government, an arm of the state, though analogous arguments can be used for planning in other sectors. There are at least four broad kinds of policy instrument: (i) expenditure — either capital or current, on equipment or wages, for goods or services; (ii) regulation or control; (iii) fiscal policy; or (iv) the form of government or planning organization itself. The efficient 'setting' of policy instruments is of major concern whatever the underlying value system.

So how do planners actually proceed? I have argued elsewhere (Wilson, 1968), basing these views on the work of authors such as Chadwick (1970), Harris (1965), and McLoughlin (1969), that there are three main kinds of activity involved in planning. Although this framework has been criticized, essentially for supporting the status quo (e.g. by Scott and Roweis, 1977), it still seems to hold and, for the kinds of reason sketched above, to be applicable in widely differing systems and situations. However, the criticisms will be elaborated and discussed further below.

The three kinds of activity are: policy, design, and analysis, which are perhaps best discussed in reverse order. They are also closely related, of course. Analysis has been the basic subject matter of this book so far: seeking as deep an understanding as possible about the structure, development, and evolution of systems of interest; and, as we noted in Chapter 1, this can provide a basis for planning. This basis can consist of a number of elements: forecasting of variables, like population levels to give a simple example, which may create new problems or demands in the future; assessing the impacts of plans — essentially, measuring the impact of new settings of policy instruments over a period of time; investigating the stability and resilience of system structures; investigating the nature of problems or objectives which have only been stated in a broad way; pointing out possibly unintended consequences of actions; the building of appropriate information systems. This list is long, and potentially longer, and involves marshalling all relevant understanding from the kinds of methods outlined earlier. A number of other features of analysis in a planning context will emerge as the discussion proceeds below.

Design is concerned with the invention of plans, but particularly with the generation of alternative plans: alternative ways of setting policy instruments to meet goals or to solve problems. The difficulty of the design process is its combinatorial nature: in the design of a road network connecting given settlements, for example, there is an enormously high number of alternatives. Then, as Steger (1965) once put it, the design problem is concerned with 'the efficient generation of efficient alternative plans.' The complexity of the combinatorial problem of design is such that it will not often be possible to use formal optimization procedures. However, there are examples where this is useful, some of which are presented in Chapter 10. Further, Harris (1981) has argued that the procedures of mathematical programming offer a paradigm for planning as a whole. In the context of design, it is useful to recall the notion of *constraints* and that one of the tasks of the designer is to find *feasible* solutions. This also connects to Simon's notion of 'satisficing'.

Finally, it is useful to recognize explicitly two particular dimensions of the design problem. Plans are often thought of as being concerned with the construction of major new facilities, but it should also be recognized that they should be concerned with the mode of operation of the facilities also. This is the distinction between, say, road design and traffic management. The designer, in the broad sense, is responsible for both tasks. At times of decreasing public expenditure, the operational plan—essentially obtaining improvements by regulation, that is shifting the emphasis from one type of policy instrument to another—becomes increasingly important.

Plans are designed to meet objectives. These objectives are set at the 'policy' stage—and this, therefore, involves many more people than those who would call themselves planners in a professional sense: politicians, administrators, voters—depending on the underlying system. The existence of goals and objectives thus provides the basis for the evaluation of alternative plans and the choice of the best plan. This problem can be stated more explicitly another way: it is necessary to measure all the impacts of alternative plans in sufficient detail to assess the incidence of the impacts on different groups. Part of the policy question is then the *weighting* which should be given to these different groups. Policy is also concerned with another kind of weighting: there will usually be many objectives, and these too must be weighted relative to each other. The whole issue of evaluation and assessment is then complicated by the often intricate relationship arising from the contribution of particular policy instruments to different objectives. We will explore this further below.

The discussion above implies that planning is a tidy, orderly business. In practice, this is unlikely. There are difficulties at each stage of the process. In analysis, there are the problems of forecasting under conditions of great uncertainty. For example, in recent years there have been substantial increases in oil prices and other corresponding changes which have made a nonsense of many forecasts. In policy-making, there is likely to be substantial conflict over weighting: the relative importance of different groups of people and different objectives at different times, particularly when resource constraints are severe. It is in any case difficult for all concerned to be explicit and articulate about all the objectives involved. The designer has to operate within all these uncertainties. Further, planning, even in relation to particular systems, if often carried out by a variety of institutions, as we will see in Section 9.5, and this of itself makes for difficulties.

These considerations mean that planning is not a tidy, linear procedure, but a cyclic one with updates and revisions continually being made at each stage. This sometimes tempts people to ask: in the face of all these difficulties is it worth it? The answer to this question will vary with the particular situation. However, it should always be borne in mind that when no formal planning is carried out, 'planning' decisions of the type outlined above are still being made, at least by implication. Forecasts are then guesses or back-of-envelope calculations, and whatever the errors associated with more formal procedures, such guesses are likely to be prone to greater error. Similarly, even if objectives are not explicit, and alternatives are not evaluated, then there are implicit objectives and choices and these may not be the ones which were intended.

9.4 SOME PARTICULAR FRAMEWORKS AND TOOLS FOR PLANNING

9.4.1 Corporate planning and related frameworks

We have seen in many contexts throughout this book that systems analysis is concerned with the consequences of interdependencies. The consequences of this concern are now reflected in many formal planning frameworks. They tend to have two things in common, whatever the name. First, a concern with overall system objectives, and secondly an explicit recognition of interdependence. The two ideas are linked. If each unit within an organization plans on its own, without full regard for other units, then the result will be inefficient because of the interdependencies, and any overall objectives will not be achieved. Hence the argument is that the process should begin the other way round. In public bodies such a procedure goes under names like PPBS (Planning-Programming-Budgeting-System), or in private and public bodies alike, corporate planning.

9.4.2 Lange–Lerner planning

One of the problems with corporate planning that it can demand 'too much' central control. What is usually meant by this is that central bureaucrats become responsible for planning decisions in relation to units about which they cannot possibly have adequate detailed knowledge. In Beer's terms, and Figure 9.1, it is as though the planners at levels 4 and 5 are directly responsible for level 1 planning and decisions. Beer's diagram, in fact, gives part of the answer: there needs to be an adequate information flow between the levels in both directions.

Marglin (1963) describes a formal way out of this dilemma offered by the Polish economists Lange and Lerner. They distinguish between 'central planning' and 'field units' and the final plan is an outcome of iteration between the two levels. In effect, the central unit ensures that overall budget constraints are satisfied for each future planning period, and that projects submitted by the field units are evaluated on a comparable basis. The field units use their local expertise to devise and to evaluate projects, in effect within constraints set by the centre in such a way that some kind of equity is guaranteed between units.

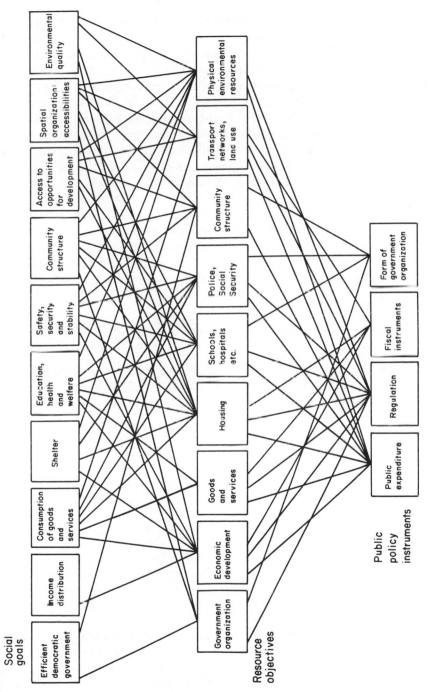

Figure 9.3 Policy instruments and social goals

9.4.3 The relation of policy instruments to goal achievement

We noted in Section 9.3 that there was a complicated relationship between policy instruments and goal achievement. Since this is a vital issue for the frameworks sketched in the preceding subsections, and since we proceed to more detail in subsequent subsections, this is an appropriate point to analyse this relationship.

The point is perhaps best illustrated by the most complicated of our three examples: cities. It is useful to distinguish between social goals, which relate to people, and resource objectives, which relate to the disposition of artefacts such as homes, schools, and so on. Then the operation of public policy instruments can be seen to relate in the main to resource objectives and these, in turn, to social goals. Such a relationship, at a broad scale, is shown in Figure 9.3, which is a more elaborate version of Figure 1.10 (from Wilson, 1974). The essential argument is then that the exercise of any one policy 'lever' can often affect a number of resource objectives, but, more commonly, any such objective is likely to be related to a number of social goals. Improved housing, for example, not only affects the comfort of living but possibly health and educational goals, and perhaps access to services of all kinds. What is being represented on this kind of figure is much of the range of interdependencies in the system set out in a way which is directly relevant to planning.

9.4.4 Cost effectiveness, goal-achievement, and cost–benefit analysis

If a project relates to a well-defined objective, then the planning task can be seen as designing the project in the most cost-effective way. For a single objective which has to be met as an absolute standard, this would involve cost minimization. The key point to note here is that the focus is on costs and on objectives not necessarily measured in money units.

Suppose now that several goals are involved and that each alternative plan has an impact on a number of different groups of people. Then, formally, we can define B_g^{nm} and C_g^{nm} as the benefits and costs associated with goal g, and group n for the mth plan. Note that benefits and costs may not be measured in money units. Then the total benefit associated with the nth plan is

$$W^m = \sum_j \sum_n \alpha_g \beta^n (B_g^{nm} - C_g^{nm}) \tag{9.1}$$

where α_g and β^n are the weights which relate to goals and person-groups respectively. This simple formulation has the advantage of making the nature of the weighting problem explicit. Of course, if B_g^{nm} and C_g^{nm} are measured in different units across goals, g, then the weights α_g will include factors which (implicitly) make the units comparable.

The task of the systems analyst in a planning context can then be seen as to use models to assess the impacts of each plan measured in goal achievement terms (and this involves all the complexities of Section 9.4.3), and manipulating policy instruments to maximize W^m. This suggests the possibility also of using formal optimization procedures, as we will discuss briefly in the next subsection.

Finally, we note that cost–benefit analysis represents an attempt to operate with a framework like that of the goals–achievement matrix but with all the costs and benefits measured in monetary units. This is particularly difficult on the social benefit side of the equation, where such concepts as consumer's surplus are used.

9.4.5 Optimization methods

We simply remark here that if a planning problem can be well defined as a mathematical problem, an optimal control problem or, say, as a branch-and-bound problem in something like network design, then the methods of Chapter 6 can be used for finding the best plan. In effect, this is the use of formal optimization methods to solve the combinatorial problem of design.

9.4.6 A note on examples

The application of systems analysis in planning comes at a late stage of the whole analytical process. In Chapter 10, we attempt an eclectic application of methods to the three examples and so it seems best to leave planning illustrations until that point.

9.5 PLANNING AND INSTITUTIONS

9.5.1 Introduction

We noted earlier that conventional planning theory of the type described above has been criticized because it takes inadequate account of its social context. It may be true, at least in some instances, that the shape and practice of planning is determined — *in a way which is not acknowledged* — by the nature of the society of which it is part. However, it takes us beyond the scope of this book to explore this here. The reader is referred to the Scott and Roweis (1977) paper and the monograph by Sayer (1976) as well as some broader criticisms of systems analysis along these lines (Berlinsky, 1976). Here, we simply observe that planning activities of the kind described are carried out in a wide variety of institutions which affect the three examples we have used as illustrations. In Section 9.5.2 we explore this topic in general terms, and in the following subsection we give a number of examples.

9.5.2 Typologies of institutions in environmental planning

Perhaps a now-standard point of systems analysis should be made at the outset: institutions are distinguished by scale, both spatial and sectoral. The former is best illustrated by government authorities which range from the international and national to the main tier of local government to even smaller community or parish councils. The sectoral scale relates to the area of activity a government department covers (and in this context, recall that 'form of government organization' is an important policy instrument). Interestingly, we usually associate 'breadth' with 'coarseness' in this context, but in the planning case it may be associated with an attempt to integrate interdependent features of planning across a field — as with the integration of the Ministries of Housing and Local Government and Transport to form the Department of the Environment. (The fact that they have now been split again perhaps indicates that there are problems of scale to set against the benefits.) These distinctions of scale could also be applied to very large companies which also have regional and functional structures.

Environmental planning as we have implicitly defined it is concerned with a large range of topics. It seems useful, therefore, to classify some approaches to planning in a general way before we proceed to discuss particular examples. A number of catagories can be defined as follows:

(i)   Resource-based (land, water, energy, and utilities).
(ii) Person-function-based (housing, shopping, transport, education, health, etc.).
(iii) Economic-function-based (industry by sector, offices, economic services like finance and insurance).
(iv) Place-based (particular places, at different scales).
(v)   Person-community-'package'-based.

The first four are self-explanatory, though it should be noted that there are some overlaps, a sense, between them. A local authority, for example, might represent a place, while a department of the authority a person-function-based service such as education. The fifth category arises from taking a systemic view of the needs of individuals, families, or perhaps communities: can the 'package' of facilities and services be adequately planned as a coordinated system? It is less easy to think of examples of planning agencies in this particular category than the rest. In a sense, it represents a coordinated look at what we called 'social goals' in Section 9.4.3.

Bennett and Chorley (1978, p. 255) make an interesting point in relation to this kind of list: that what we called n-groups in relation to the goal achievements matrices, B_g^{nm} and C_g^{nm}, can be further subdivided in this way. In other words, that 'distributive information' can relate to clients, sectors, and areas, which are like categories (ii)–(iv) above. They also note, following Gulick and Urwick (1937) and Self (1972), that the principles of departmentalization by government agencies can be related to similar categories — in this case: client, purpose (for each control goal, 'favoured in the United Kingdom since Haldane in 1918' for government department structure), process (type of expenditure or skill), and area. It is appropriate not simply to note their particular categories, but the more general point that forms of planning and governmental organization can be related to categories of this or other types. Usually one has to be chosen on the basis of the main form of structure, but allowance may then have to be made for the 'demands' of other categories.

9.5.3 Some examples

We consider examples of planning systems in each of the categories introduced in the previous subsection.

Resource-based planning systems

Some resource-based systems are relatively simple in overall structure. In the case of water, for example, there are ten Regional Water Authorities covering the whole of Britain and a National Water Council with an overseeing, monitoring role. The Department of the Environment is the corresponding central government ministry. A more complex example is provided by land-use planning, and we concentrate on that in the rest of the section.

The heart of the land-use control system in Britain is development control which mainly operates passively in that it responds to applicants. The more positive side is the production of development plans which form the basis of development control policies. Both of these functions are carried out by local authorities, though the picture has become complicated since the 1974 reorganization of local government because, in the case of metropolitan areas, the planning work is divided between the County (Structure plans—essentially strategic plans for the whole county) and Districts (District and local plans— more detailed than the Structure plan but conforming to it). These plans have to be approved at central government level by the Secretary of State for the Environment, who is also responsible for appeals against development control decisions. (The appeals procedure might be considered as an example of the analogue of the sympathetic or parasympathetic nervous systems in Beer's schema.) The public can be involved through planning inquiries on major plans or developments which are also organized by the Department of the Environment, or through public participation exercises at the local level.

There are other minor complications: the Regional Economic Planning Councils, which cover a bigger area than counties, have often produced plans which have strong land-use implications; and in one or two rural cases, the planning authority is not the local authority, but the corresponding National Park Planning Board.

It is instructive to review briefly the features of this organization in the light of some of the other concepts introduced in this chapter. The variety of the control system is essentially the amount of knowledge the planners can accumulate on this area and its development processes. This is then to some extent recorded in the plans and partly in any research and intelligence team which back them up. There is also the variety associated with, for example, planning committees who bring a different kind of knowledge as representatives of the public to the decisions they make.

In terms of Beer's schema, activities $A, B, C, \ldots$ may be taken as the final implementation of (or agreement of) projects on the ground. The levels 3–2–1 management systems would represent development control and corresponding machinery. Level 5 would be the higher level policy making of planning committee (and council) and officials. Level 4 would be the intelligence operation and perhaps include the plan-making functions. A broadly similar interpretation in these terms is offered by McLoughlin (1973), as shown in Figure 9.4.

The massive complexity of land-use planning arises from the fact that most people, organizations, or activities in cities (say) are consumers of land. This generates many conflicting objectives and the conflicts have to be resolved in a context of imperfect understanding of the costs and benefits of alternative proposals. This also, of course, generates a lot of interaction with other kinds of

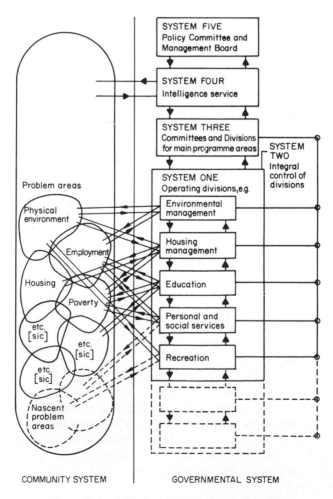

Figure 9.4 A CNS-system applied to planning

institutions. And, of course, the combinatorial problem works here with a vengeance: there is always a very large number of possible alternatives. So, while some of the formal methods of Section 9.4 have been widely applied to specific sectors within a plan—especially to transport—they are less developed for land-use planning as a whole. Exceptions are Lichfield's (1969, 1970) use of the 'planning balance sheet', which is a form of goals–achievement matrix applied to land use, and some limited applications of mathematical programming which will be discussed in Chapter 10.

Person-function-based planning systems

As an example of a service which affects everybody but which is separate from the main agencies of (at least, local) government, consider health. The central government ministry involved in Britain is the Department of Health and Social Security. It is at the top of a pyramid-hierarchy: the Department allocates resources to *Regional Health Authorities* (RHA's). Within each of the regions are a number of *Area Health Authorities* (AHA's). The allocation of resources to these, and the monitoring of performance, is by the Regions. Each Area Authority is divided into a number of districts—usually two or three, but it can, exceptionally, be only one—and the health services in each are the responsibility of a *District Management Team* (DMT). Each DMT reports to its AHA.

The RHA's and the AHA's are each public authorities. At the regional level, members are appointed by the Secretary of State; at the area level, the chairman is appointed in the same way, but the members are appointed by the RHA. Many of the appointments, at each level, are on the nomination of various bodies, ranging from local authorities and trade unions to the various professions involved in the service. This ensures a wide range of public representation, but there are no elected members, except indirectly through the local authority representatives. The final element in the formal structure consists of *Community Health Councils* (CHC's). There is one of these for each district. They are intended to represent the public in a more direct way. They have an office in each district, they actively seek consumer viewpoints, and their own voice is heard through representation on the appropriate AHA.

An illustration of the formal system, for a typical Regional Health Authority, is shown in Figure 9.5. There are planning functions at each of the spatial scales: region, area, and district. Formal plans are produced at each scale, for three years ahead with the District plans being updated annually. The influence of the corporate planning idea can be seen on both management and planning structures. The management terms are made up of representatives of the main professions—medical, nursing, financial, and administrative—and they operate in a corporate way in relation to services broken down under a number of main headings. These include, for example: acute (the major general hospitals), children (including maternity), the elderly (including geriatric hospitals), the mentally ill, primary care, and so on. There is also the task of liaising with the local authority on the joint provision of some services—for the elderly and

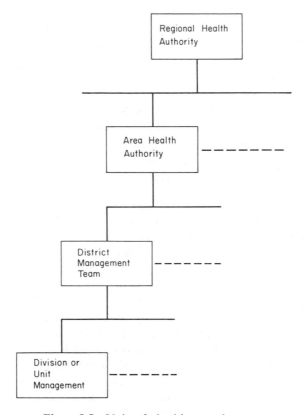

Figure 9.5 Units of a health control system

physically and mentally handicapped, for example—and there is a Joint Consultative Committee made up of members of the local authority's social services committee and the AHA. There are obvious interdependencies here: if the local authority is forced to cut down on care for the elderly at home, there can be a substantial increase in demand for hospital beds and the overall cost of a 'cut' in expenditure can exceed that of the original situation.

How can we assess this kind of structure in relation to our earlier concepts? We begin with an observation on requisite variety and then discuss a new idea in relation to Beer's control concepts.

The DMT is the top of the 'line management' pyramid. However, the units are often large and complicated, like a large hospital, and it is unlikely that a DMT will have sufficient variety to exercise detailed control over each unit. They will, of course, exercise a lot of control, but what they can also attempt is the monitoring of certain key indicators for each unit—like waiting lists for different kinds of hospital service.

In terms of Beer's CNS system, these examples illustrate hierarchical control. Let us consider the extreme first. At the DHSS scale, the output activities $(A, B, C, \ldots)$ are those of the Regional Health Authorities and so planning and monitoring indicators will relate mainly to that scale. Level 5 will be the ministerial and higher civil service level, subject to the control of Parliament (which can be taken as part of level 5 or as part of the 'environment' depending on the viewpoint to be adopted) and subject also, of course, to Cabinet control. Levels 3 and 2 will be the civil service management and coordination systems and level 1 will relate to each RHA. This, as we will see, can be expanded into a Beer-system on its own, as is typical of hierarchical systems. Level 4 will include subsystems like the Operational Research Division, but this may well be relatively rudimentary in relation to the degree of control which is required. The sympathetic and parasympathetic systems will function when Ministers are alerted to particular incidents from grass-roots levels, possibly, for example, by back-bench M.P.'s.

At the hospital level, the activities $A, B, C, \ldots$ will be the major functions and back-up services. Level 5 will be the unit management team and there will be identifiable structures at levels 3 and 2. Again, one suspects that level 4 is in a rudimentary state, probably with its functions absorbed, possibly inefficiently, by level 5 personnel.

There will then be other Beer-like structures at Regional, Area and DMT scales.

Finally, we note that the task of getting the relationships right between hierarchical levels has many of the features of Lange–Lerner planning. A balance has to be struck between the demands for control in relation to efficiency and equity from higher levels, and the superior knowledge of particular units and services which exists at lower levels. This is generally a fascinating and difficult problem for hierarchical control systems.

The overall health service control system is very complex. There is almost certainly an inadequate control linkage between the Area and DMT levels, and this is particularly important because the Area is a 'public' level, the District not. The government (Department of Health and Social Security, 1979) is proposing to rectify this kind of deficiency by combining these two levels into District Health Authorities (DHA's).

Economic-function planning systems

These perhaps concern us least in this book because we are mainly concerned with the public aspects of planning environmental systems. Most major economic sectors are private or, even in the case of nationalized industries, operate with similar kinds of management and planning structures. The main distinguishing feature from other service systems we have considered is that they usually have marketable products and have to make profits, break even, or achieve financial targets. This means that financial accounting dominates much of the planning systems of such enterprises. It is also in general true that stronger control can be exercised in parts of the system because there is no direct public accountability. However, in many cases, there are complicated relations with government and with other organizations such as trade unions.

Place-based planning systems

We noted earlier that the main organs of government, at different spatial scales, are the main place-based systems. The interests of a 'place' can be seen to be uppermost in the 'minds' of local authorities which are competing with each other to attract firms to provide more local employment. The exercise of central government control is immensely complicated relative to, say, a single-service sector like health, because so many different Ministries are involved. The overall allocation of resources is the responsibility of the Department of the Environment through the rate support grant, but many other Departments — Education being an obvious example — have close relationships. Overall, however, we can observe that there will be a hierarchical control system of the same type, but more complicated, as that sketched for the health service, but further analysis would take us beyond the scope of this book.

Person-community-package-based systems

We observed earlier that there were no agencies with this particular concern, though it could be argued that the monitoring of the range-of-choice (degrees of freedom?) or 'packages' available to different kinds of individuals would be a useful role for local planning departments and that the kind of micro-simulation model outlined in Chapter 8 would be a good basis for such work from a systems-analytical viewpoint.

9.6 SUMMARY: THE ELEMENTS OF PLANNING AND CONTROL

It would have been possible in principle to write one or more whole books on the topics of this chapter. The result may seem inadequate in a number of ways. It should be stressed, therefore, that what has been attempted here is to highlight a number of features of planning and control which seem to be common to a wide range of problems and approaches. There has been an emphasis on the nature of control processes within planning (and, by implication, associated implementation and management) systems. At a broad level, the systems analyst needs to be aware of requisite variety and of the concepts introduced by Beer. Much of the rest of the work of control system design then has to be carried out, *ad hoc*, in relation to particular systems because there is no general theory. Similarly with planning: there are different kinds of activity at a broad level — like policy, design, and analysis — and a number of basic tools. An awareness is needed of these, and again much detailed work has to be carried out in relation to particular examples. These notions will be illustrated in more detail in relation to our final synthesizing look at the three main examples of the book in the next chapter.

NOTES ON FURTHER READING

Two books which convey a useful account of urban planning from a systems viewpoint are those by †McLoughlin (1969) and †Chadwick (1971, Second

Edition, 1978). The broad background of this 'movement' in planning is interestingly described by †Hall (1975, Chapter 10).

The remaining references refer mainly to specific subsections. The law of requisite variety is described by *Ashby (1956, Chapter 11) (9.2.2) and the control system of Section 9.2.3 in †Beer (1972, Second Edition, 1980a; 1980b). *McLoughlin (1973) is also useful background reading in both contexts. The policy-design-analysis framework for planning is described in †Wilson (1974, Chapter 2). Some useful 'corporate planning' references are by *Stewart (1971, 1974),*Eddison (1973), and *Greenwood and Stewart (1975) (9.4.1). See also Clarke (1978). The account of Lange–Lerner planning is based on the description by *Marglin (1963, Chapter 5) (9.4.2). The relationship of policy instruments to goal achievement is described in *Wilson (1974, Chapter 2) (9.4.3). *Chapter 13 of the same book contains reviews of a variety of evaluation methods up to and including cost–benefit analysis, and other references are cited there. †Howe (1976) is a good review of cost–benefit analysis in the water resource context, and the †Yorkshire and Humberside Economic Planning Board (1976) report is a useful description of similar issues for moorland areas. A recent description of 'great planning disasters' may also be timely in this context—see †Hall (1980) (9.4.4). The combinatorial problem of design is described in different ways by *Scott (1971) and *Alexander (1965). A view of the planning process based on an optimization paradigm is presented by *Harris (1981).

PART 3

Examples of Environmental Systems Analysis

CHAPTER 10

Applications and conclusions

10.1 INTRODUCTION

In Chapter 1, we summarized the basic objectives of systems analysis as being concerned with (i) handling complexity, (ii) identifying and understanding systemic effects, (iii) identifying methods which can be applied generally to particular system types, and (iv) providing tools which are aids to planning. These ideas were illustrated in relation to moorland ecosystems, water resource systems, and cities. The basic concepts of systems analysis were introduced in a qualitative way in Chapter 2, and in Chapter 3 the 'methods perspective' to be adopted was explained. A wide range of methods was then described in Chapters 4–9. The applications of these methods were illustrated within each chapter, mainly using the three examples of systems introduced in Chapter 1.

It is now appropriate to attempt a synthesis for each of the examples. Indeed, it is in the spirit of systems analysis that we should do so. In the next three sections, therefore, we take each of the examples in turn, and explore the task of building a more complete model, assembling eclectically whatever set of methods seems to be appropriate in each case. We can then measure what has been, and what can be, achieved against the objectives of systems analysis as recapitulated above. Unfortunately, the results are inadequate: the intellectual programme of systems analysis has not yet been completed in relation to these examples. So as well as, implicitly, assessing the state of the art, we also try to point out directions of future research.

Finally, in Section 10.5, some concluding comments are made.

10.2 MOORLAND ECOSYSTEMS

10.2.1 Introduction

The task of this subsection is to attempt a comprehensive systems analysis of a moorland ecosystem using whatever methods seem appropriate. It is implied in Chapter 2 that the first task is to identify the components of the system, at appropriate levels of resolution, and their interdependencies, together with the basic processes which govern the system's dynamics. The other concepts introduced in Chapter 2 all bear on various aspects of these features. We also outlined principles for constructing system diagrams. These considerations are all explored in Section 10.2.2.

Entropy-maximizing methods (Chapter 4) have no direct relevance for moorland ecosystems; nor do optimization methods (Chapter 6)—except in a minor way in connection with planning to be taken up below, and at scales which are not our central concern; nor do network analysis methods (Chapter 7)—again except at other scales, for example in the tree-like structure of particular plants or in digraph analysis which was dealt with directly in Chapter 7. Accounting relations are important, but the principles have been presented in Chapter 5. We use these in the main subsections below on dynamics (10.2.3, following the methods of Chapter 8) and on planning and policy (10.2.4, following the methods of Chapter 9).

10.2.2 Description and basic concepts

The main features of moorland ecosystems were presented in diagrams such as 1.1 and 7.24, and the main processes involved at this scale were outlined in Chapter 1. They can be summarized very briefly as: plants fixing energy by photosynthesis, drawing on sunlight and water from the atmosphere and nutrients from the soil; herbivores, like sheep, eating plants; carnivores eating animals; decomposers returning nutrients from dead matter and faeces to the soil. In Figure 1.2, we gave a more detailed account, following Gimingham (1972), of the particular components, listing them in the form of a food web. And this, of course, represents one way of representing interdependencies: in this case, the prey–predator relationships. The second main form of interdependence arises out of competition for resources. This is of two kinds: competition for basic resources, as plants competing for light, water, and nutrients; and competition between predators who share at least one common prey.

We conclude this subsection by noting some of the main kinds of question we would like to answer in this kind of system analysis. The detailed issues of description and level of resolution are dealt with in the context of constructing a dynamic model in the next subsection. According to Chapter 2, we should seek to establish whether ecosystems achieve equilibrium, or what other system-behaviour patterns arise. Experience suggests that there are many answers to this question. On different moorlands, different species may dominate. The most obvious ones are heather or bracken. In the case of the former, if the moor has been untouched, there will be a variety of ages of heather, and this would represent a kind of steady-state equilibrium. But when and why does one species dominate? Under what circumstances can there be an invasion or succession to dominance by another species? These are the questions we try to answer with a general formulation of a dynamic model, based on Chapter 8, in the next subsection. And we should also note that the evident variety of states is in part generalized by a major omission from all our diagrams so far: the influence of people in farming and management policies turns out to be crucial. We build these features into the dynamic model and then consider the implications for planning and policy in Section 10.2.4.

10.2.3 Aspects of ecosystem dynamics

The foundations of dynamical analysis for ecosystems in general were laid in Sections 5.2.3 and 5.4.6. The essential basis was the definition of compartments and the specification of flows between them. The stocks and flows are measured in terms of biomass or energy units, and occasionally 'numbers' of animals or plants are added.

At first sight, it seems that the main task for modelling a moorland ecosystem is simply to specify the species for each compartment—definition of the sectoral scale; but it turns out that the resolution decisions involved are potentially quite complicated because of the need to get an adequate representation of the processes involved. For example, as we saw in Chapter 1, the amount of cover provided by heather is very much a function of age: small for young heather, almost total in the building stage, and progressing to small again for 25–30 year old (degenerate) heather. This obviously has a major effect on competition for light with other plant species and so suggests that 'age' should be distinguished. Another important process at present is invasion of heather by bracken (Smith, 1977a, b, c), and this is a spatial process: the bracken 'frontier' moves under the ground. This suggests the need to distinguish location so that the effect of the presence of different species nearby can be built into the model.

The basis for our analysis has to be something like equation (5.90), which is repeated here for convenience:

$$S_n^k(t + \delta t) = \sum_m \phi_{mn}^k S_m^k(t) + \phi_{en}^k E^k. \tag{10.1}$$

The subscripts m and n label compartments and k is the 'quantity' in each compartment represented by this equation—it could be total biomass or it could be a particular nutrient. For convenience, let us assume that we can label one or more compartments to represent the environment, and that these are numbered within the m, n subscripts. The equation would then become, more simply in notation but without loss of generality,

$$S_n^k(t + \delta t) = \sum_m \phi_{mn}^k S_m^k(t). \tag{10.2}$$

By redefining the coefficients ϕ_{mn}^k—because recall from the discussion in Section 5.4.6 that they can be complicated and involve S-terms—an equation like (10.2) can be written in differential equation form as

$$\dot{S}_n^k = \sum_m \phi_{mn}^k S_m^k. \tag{10.3}$$

We can redefine the ϕ_{mn}^k yet again and write the equation in quasi-logistic

236

form (and one which distinguishes processes like birth or sometimes death which are 'internal' to a compartment from those which are interactions):

$$\dot{S}_n^k = S_n^k(a^k - \sum_m b_{nm}^k S_m^k).$$ (10.4)

The first important point to recognize in this general formulation is that the coefficients can, in general, be of either sign. For example, a prey population would have positive a_m^k and at least one positive b_{mn}^k; a predator population would have negative a_m^k and at least one negative b_{mn}^k. Competition between species m and n would be recognized by negative b_{mn}^k and b_{nm}^k in the m-equation and the n-equation respectively. Recall also that the coefficients a_m^k and b_{nm}^k can take a great variety of functional forms, as discussed in Section 5.4.6, and it can be seen that, potentially a system of equations such as (10.4) can generate a very rich variety of system behaviour.

The next step is to recall the argument that we need to define compartments, at least for some species, which distinguish age and location. This means, in effect, taking labels like m and n to be subscript lists. A compartment would be defined by, say, (m, r, i) for (species, age, location) and if we replaced n by (n, s, j) on a similar basis, equations (10.4) could be written

$$S_{nsj}^k = S_{nsj}^k(\sum_{s'j'} a_{ns'j'}^k - \sum_{mri} b_{mri,nsj}^k S_{mri}^k).$$ (10.5)

This is potentially unwieldy, of course, particularly the seven-dimensional coefficient $b_{mri,nsj}^k$. Note that there is also now an additional summation: over other age groups and locations in relation to what was the a_n^k term. However, the saving grace is that there will be relatively few interactions between locations— only contiguous ones, and only ones where 'invasion' conditions are satisfied. The age-group information would be built into the functional forms of the b-coefficients. And finally, the number of interspecies interactions, while large enough to generate interesting systemic behaviour, are still relatively small. However, there do not seem to be any models in the literature which have been developed at this degree of detail. Indeed, it is difficult to develop realistic models at all because of the problem of assembling suitable data to estimate the functional forms of the coefficients in equation (10.5) and the parameters associated with them.

The next step in the argument, therefore, must be an essentially theoretical one: to explore how various biological phenomena and information of relevance to ecosystem development could be built into models of the form of (10.5), and to consider the kinds of behaviour which solutions to the resulting equations might represent. We follow mainly Gimingham (1972, 1975) and Smith (1977a, b) in relation to the first topic, biological information, and Maynard Smith (174) in relation to the second, the nature of the solutions. We mix into the argument some discussion of various model-design decisions which have to be taken.

The first major decision relates to the identification of species. It is clear from the food-web diagram from Chapter 1 (Figure 1.2) that a large number of animal and plant species could usefully be defined. We should obviously be particularly concerned with dominant or potentially dominant species such as heather and bracken (and possibly birch or other trees in certain situations) and it may be possible to lump most of the rest together. A particularly interesting point is the probable need to identify, say, bracken and birch, even though they may be practically invisible on a moor dominated by young heather, because of the possibility of invasion or succession at the degenerate stage.

The second decision relates to the nature of the use of the moor by human agents. It is clear that the current dominant form of a moorland ecosystem is often more determined by past and present human practices than by anything else. There are three major types: (i) the clearance of woodland (or other vegetation), which is thought to have been the basis of the creation of the present heaths, dating back to neolithic times; (ii) the use of the moor for grazing sheep or cattle, or as grouse moors—and it is grazing which, typically, is thought to have prevented any resurgence of tree growth on most moors; and (iii) management practices such as burning, which is very common, or perhaps ploughing or the control of bracken by chemical means. The model-design decision which relates to this because the model must reflect the practices which have been imposed on the moor in the past and in the present. If grazing is a major use, then sheep and any other grazing animals must be included as compartments.

The third broad class of model-design decisions relate to the terms which appear as coefficients in equations (10.5) which may in practice take quite complicated functional forms. It is here that the detail of the biology has to be built into the model, as we will see.

To fix ideas, let us concentrate mainly on heather or bracken as dominant species—in competition with each other—and on sheep, cattle, and grouse as the feeding animals. The latter are all herbivores, of course, and are not themselves controlled by predators—unless you argued that human agents were the predators of sheep and cattle: but such control would be known as 'management'! We also consider only the most obvious management policies initially: burning, and the control of grazing. What, then, are the biological processes to be represented in the coefficients of equations (10.5)?

We saw in Chapter 1 that the age of heather is crucial in determining the amount of cover it provides. The coefficients which represent competition between it and other plants must therefore include this. There is an additional complication: that the shape of the heather is determined not only by age, but also by intensity of grazing (Gimingham, 1975, p. 33). The competition is partly due to shade, but is also due to 'root competition' expressed in terms of access to nutrients.

The other main effects arise out of interaction with other system elements. For example, the use of the moor for grazing is a contribution to nutrient depletion— via the carcases of animals which are transported away. Also, the effect of control of heather by either grazing or burning to keep it relatively 'young' and therefore with high cover, is to generate a 'monoculture'. This weakens the diversity of the ecosystem (and possibly also weakens its stability and makes it more prone to invasion). This process is also known to encourage soil deterioration (podsolization).

While heather can support sheep, grouse, and occasionally cattle, bracken is considered as an infestation. It is basically poisonous to animals and so a bracken-covered moor is unproductive. In recent years, it has spread rapidly in many British moorlands and so it provides a good example of invasion, with associated problems. However, there is no total agreement on the mechanisms involved and here we only sketch some basic ideas, following Smith (1977a).

The first point to note is that while heather itself spreads by seeding (and with some seeds germinating after the winter, some before, to improve survival prospects), bracken cannot do this in typical moorland conditions. It spreads through extensions of the rhizome system under the soil. This makes spatial contiguity particularly important.

Present evidence seems to be that its competitive position is improved by many present management practices, such as intensive grazing and burning. (The rhizome system is practically immune to fire, for example.) This is either because, with its strong root system, it is in a good position to invade after burning or grazing of young heather, or because the management policies ultimately affect the soil in a way which favours it. It is also interesting to note that, in principle, the coefficients which represent growth should take account of the way plants develop during a season: bracken develops late, and its more successful competitors such as some grasses develop early, before extensive bracken cover develops.

We need a model, therefore, which will explain the main features of stability and diversity, invasion and succession, in particular climatic and soil areas in relation to the management policies which are pursued. First, therefore, we offer a model which includes many of the essential features to provide a basis for progress; later, we elaborate it.

We built, first, a six-compartment model: heather (H), bracken (B), other vegetation (V), sheep (S), cattle (C), and grouse (G). The human controllers (P) provide, in effect, the upper level predator but we will include their effects directly; there is no additional benefit to be gained from the formal definition of a seventh compartment—except perhaps in the system diagram in Figure 10.1.

This shows that the controllers 'prey on' the three animal types. Sheep and cattle eat heather and vegetation; grouse eat only heather. Nothing eats the bracken. However, the animals are competing for resources in the form of food, as are the plants for resources in the form of light and nutrients.

In building the model, we follow the methods of Maynard Smith (1974, p. 108) and it is convenient to follow his notation:

H : 1
B : 2
V : 3
S : 4
C : 5
G : 6

and to let X_i be the biomass of compartment i. He works in difference equation

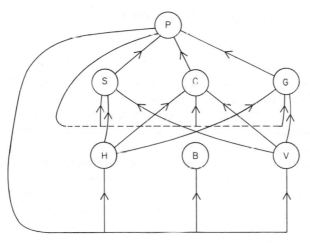

Figure 10.1 Linked compartments for an ecosystem model

form; we will use differential equations. Our assumptions so far can be represented as follows:

$$\dot{X}_1 = X_1(a_{10} - a_{11}X_1 - a_{12}X_2 - a_{13}X_3 - a_{14}X_4$$
$$- a_{15}X_5 - a_{16}X_6) \tag{10.6}$$
$$\dot{X}_2 = X_2(a_{20} - a_{21}X_1 - a_{22}X_2 - a_{23}X_3 - a_{25}X_5) \tag{10.7}$$
$$\dot{X}_3 = X_3(a_{30} - a_{31}X_1 - a_{32}X_2 - a_{33}X_3 - a_{34}X_4 - a_{35}X_5) \tag{10.8}$$
$$\dot{X}_4 = X_4(a_{40} + a_{41}X_1 + a_{43}X_3) \tag{10.9}$$
$$\dot{X}_5 = X_5(a_{50} + a_{51}X_1 + a_{53}X_3) \tag{10.10}$$
$$\dot{X}_6 = X_6(-a_{60} + a_{61}X_1). \tag{10.11}$$

This is a mixed prey–predator/competition-for-resources model. The self-inhibition terms in the first three equations show that vegetation growth is resource limited, and that growth rates are controlled by competition with other plants and by feeding. Competition between the different predator animals is implicit in the utilization of a common food supply. The growth of the predator animals is limited by food supply.

In prey–predator models, terms like a_{20} are usually shown as positive for prey populations and negative for predator populations, reflecting 'birth' and 'death' processes respectively. Conversely, the remaining terms in the first three (prey) equations are negative to reflect decline due to competition or predation; the remaining term in the last three equations is positive, representing birth and growth processes through food supply. We have mainly followed these conventions in the signs shown. However, a_{40} and a_{50} are shown as positive because they are partly, and probably mostly, determined by the grazing policy of human agents. Since, typically, they will be absolute additions or deletions from stocks, then suitable composite terms might be

$$a_{m0} = -a'_{m0} + \frac{a''_m}{X_m}, \quad m = 4, 5 \tag{10.12}$$

for suitable constants a'_{m0} and a''_{m0}. a'_{m0} is then the conventional death rate and a''_{m0} is the control total per unit time. Normally, of course, additions or deletions will come as one 'stock' to the system an can be dealt with either in a simulation (as we will assume here) or by reverting to a difference equation formulation.

We argued earlier that the competitive influence of vegetation, particularly heather, is a function of age. We show later, in an extension to this model, how to introduce age explicitly. For the present, we take biomass to be an indicator of competition. This implies that terms like a_{21} and a_{31} are functions of X_1 which take the form sketched in Figure 10.2. A similar argument could be used for the other plant-competition terms. The rest of the biology of growth is built into the relative scaling of the competition coefficients both between each other and compared with the terms a_{10}, a_{20}, and a_{30}. The relative magnitude of these, competition effects aside, will depend on plant characteristics, climate, and soil properties. As we saw, there are also short and long term feedbacks in this respect. The growth of bracken or trees provides shelter and changes the microclimate. Monoculture of the heather leads to podsolization of the soil; and so on. Such effects could be built into a simulation model if desired.

The animal feeding coefficients are essentially physiologically determined. The relative magnitude of these for different species measures the degree of competition.

Note that a term $-a_{25}X_5$ has been added to the bracken-equation (10.7). This is not because cattle eat bracken, but they damage it by trampling because they are more indiscriminate grazers than sheep. This is important ecologically for the control of bracken and is a useful management tool (Smith, 1977a).

The next step in the argument is to see whether these equations—even as an approximate model—could generate the types of solution which describe moorlands as we see them and invasion and succession processes.

We can learn something initially by considering typical results associated with competition and prey–predator models. In the first case, there is at most one equilibrium solution in which all the species involved coexist, but there is also the

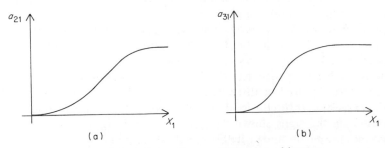

Figure 10.2 Effects of competition versus biomass

possibility of one or more of the species being eliminated. This is likely to depend on how close the 'bundles of resources' needed for each of two species are: if they are very close, the more efficient species will win out; if less so, coexistence is possible. On this basis, some 'other vegetation' will always coexist with either heather or bracken, but heather and bracken are unlikely to coexist with each other for long periods because of the competition for light. This effect will be manifested in the equations by competition coefficients increasing with biomass: if both heather and bracken biomasses increase for a stand, a bifurcation point will be reached beyond which one will become dominant (determined by the particular relative magnitudes of the coefficients in each case). We should also note that the return to equilibrium in competitive situations is not oscillatory.

A pure prey–predator system is also likely to have a stable equilibrium, but with a return to equilibrium which is oscillatory. In a combined model, therefore, with competition at each trophic level, the overall behaviour could be quite complicated.

What are the prospects for achieving long-run stable equilibrium? For moorland systems in general, the answer must be: not high. This partly depends on the long life-cycle of the heather. In an uncontrolled system, the lack of competitive edge of heather in its degenerate phase means that invasion and succession by bracken or other vegetation (grass or woodland) is always possible. On a controlled stand, the use of burning or grazing methods in effect maintain the moor in a certain kind of disequilibrium — approximately, states of maximum growth. Because of the longer run ecological effects we have discussed above, it may be very difficult even in such a situation to resist invasion by bracken without resorting to more expensive methods such as pesticides. In other words, the only situations which are likely to represent climax states and to stay in equilibrium are either bracken-dominance or, more rarely, succession to woodland. Once such states are achieved, it is easy to see how they would be maintained even in our simple model, simply because most of the competition would be eliminated.

The final step in the argument, therefore, is to consider whether the model tells us enough about transient states (which may be maintained by human control) or progression to equilibrium states. Probably all that can be said is that a great variety of behaviour is possible. There will be a number of underlying equilibrium positions and the one which is dominant at particular times may be determined by small fluctuations in the neighbourhood of critical parameters values. The presence of prey–predator relations will introduce fluctuations into transient behaviour. Since the populations of sheep and cattle can be controlled, the oscillations will mainly apply to vegetation biomass and possibly to the grouse populations.

The three kinds of vegetation have life-cycles operating on very different time scales: relatively long for heather; shorter for bracken once it is within invasion 'distance'; short for some other vegetation; much longer for succession to woodland. The interaction of these will be rather complex.

It is difficult, therefore, to make progress analytically. What is necessary is to carry out a wide range of computer simulations, varying over different assumptions about the functional form of the coefficients and the numerical values of their associated parameters; with respect to different initial conditions, since there will obviously have an enormous impact on a particular development; and finally with respect to alternative management policies.

10.2.4 Planning and control

We can both extend our notion of a moorland ecosystem and discuss planning and control through a valuable study of the Pennine Uplands carried out by the Yorkshire and Humberside Economic Planning Board (1976). They begin their Chapter 1 with a sentence which is a useful summary:

> 'The Pennine Uplands perform a triple role: they offer the local community a pleasant place in which to live, work and bring up their children; they offer visitors an attractive environment in which to relax; and to the nation they supply food, wool, water and minerals.'

This broadens our initial concept of the moorland ecosystem, mainly by the inclusion of the local community and tourists—and hence manufacturing and other population based activities, and the effects of tourism—and by distinguishing the role of the moors in water resources and mining and quarrying (and—another addition—forestry). Implicitly, the goals of any planning exercise consist of improving the income and environment of the local community, facilities for tourism (which also contributes to the first goal), and maximizing, subject to relevant constraints, the economic benefits of the associated productive activities.

Throughout the book, we have placed heavy emphasis on interdependence, and this is shown explicitly for this extended view of moorland ecosystems in Figure 10.3 which is taken from the report. This shows the matrix of main interactions between the different system elements (at a fairly coarse level of sectoral aggregation). The results are clearly set out and are self-explanatory. This matrix can be taken as equivalent to a signed digraph (c.f. Chapter 7), showing the presence and direction of interactions but not their magnitude or functional form.

The ecosystem model outlined in the previous subsection relates mainly to the elements of the system which refer to changing habitats and environments as a consequence of farming practices. In principle, the model could be extended by the addition of compartments, and by the incorporation of input–output accounts to represent the economic sectors (c.f. Chaprer 5). However, this would be an enormous task in relation to the scale of the problem and it is perhaps better to proceed directly to a consideration of planning issues based on another table from the report.

This is shown as Figure 10.4. It is particularly interesting in relation to the discussion in Chapter 9 because the column on 'opportunity' is in effect a catalogue of alternative actions, and so represents the design phase of planning (in relation to the overall implicit policy objectives mentioned earlier), and the third column lists the variety of institutions involved.

The rest of the report consist of a detailed qualitative analysis of these proposals but ends without any very conclusive recommendations. Would a more quantitative analysis be useful in this case? Perhaps the obvious section of Chapter 9 to turn to first is cost–benefit analysis. What is needed is a specification of the costs and benefits of the different proposals together with an awareness of the major interdependencies. If the benefit structure in respect of some proposals was suitably simple, they could be analysed on a cost-effectiveness basis. Let us consider a number of examples.

First, if farm amalgamation was encouraged, this would generate more effective use of labour and possibilities for investment—for example, better control of moorland grazing. The ecosystem model of the previous subsection would be useful for determining the ecological consequences of different alternatives. A decision to proceed could be taken purely on a cost-effectiveness basis.

Secondly, consider developments in the use of moorland for water resources. This mainly, at present, involves the exclusion of agriculture from Water Authority catchment areas to avoid pollution. The financial benefits to the Water Authority of extending such areas could be calculated and the decision to proceed could be taken partly on the basis of the rate-of-return which resulted, but partly on whether this rate-of-return was greater than the agricultural alternative. Thus proposals arising from different sectors could be compared.

Thirdly, and perhaps more interestingly, consider the possible multiple use of water catchment areas. There would be the possibility of adding agricultural use to such areas if the additional investment in new treatment facilities was less than the benefit of agricultural use. In this case, there is the added complexity of different agencies being involved: even if the calculation suggested implementation of multiple use, the farmers would have to be persuaded to pay for the treatment plants, or some public body would have to act as mediator. There is the further multiple-use potential of possible recreational use of water, and the problem then gets nearer to demanding a full cost–benefit analysis.

The other kind of development which is competitive in an ecological sense is forestry. Again, financial calculations can be carried out by forestry groups and decisions taken to go ahead if land-owners are prepared to lease or sell at an appropriate price. But again there are multiple-use potentials: forests have recreational uses and can also be used as shelter belts as an aid to agriculture. This last benefit may be an example of something which is difficult to quantify.

It would be possible to set out problems of this type more formally—as mathematical programming problems or as control theory problems. However, as implied earlier, the effort in this case would probably not be worthwhile. What is needed is the costing of a wide range of alternatives and an awareness of the interdependencies on the lines sketched in the examples above.

Figure 10.3 Main interactions in a moorland ecosystem

ON → / EFFECT OF: ↓	PRODUCTION				
	Farming	Forestry	Quarrying, mining, and associated industries	Water supply	Manufacturing and other activities
PRODUCTION — Farming		− Competes for land with some types of forestry		− Pollution	+ Associated industries
Forestry	− Forestry competes for land + Can be complementary through added income + Shelter belts and windbreaks		+ Can improve landscape around quarries + Can contribute to environmental restoration	+ Improves water quality	+ Encourages associated industries
Quarrying, mining, and associated industries	− Some competition for land − Dust nuisance − Quarry traffic damages walls			− Disturbance of hydrology − Pollution	+ Associated industry e.g. brickmaking
Water supply	− Competes for land − Restricts intensity of stocking − Restricts fertilizer use	+ Can produce afforestation areas	− Restriction on some exploitation		
Manufacturing and other activities	− Competes for labour	− Air pollution − Competes for labour	− Competes for labour	− Pollution risk	− Competition for labour between manufacturing industries
TOURISM/RECREATION — Shooting and fishing	− Restricts intensity of use + Added revenue	− Groose moors may compete with forestry			
Tourism	+ Added revenue				
Recreation (including educational visits)	+ Added revenue − Nuisance of visitors	+ Additional justification for planting + Added revenue + Increased job opportunities − Forest fires − Litter and vandalism			+ Souvenirs
LOCAL COMMUNITY				− Limits mining	
ENVIRONMENT — Landscape	− Planning controls can limit new buildings	− Some restriction of design, scale, and type of planting	− Could restrict expansion + Some screening of plant − Opposition to road widening for lorry traffic	− Could restrict choice of site	− Location of site, size, type of building
Nature conservation				+ Reduces soil erosion	

Note: This figure shows the interactions between the main activities in the Uplands. A positive sign (+) indicates that the interaction is generally beneficial, and a negative sign (−) that the interaction can be harmful.

TOURISM/ RECREATION		LOCAL COMMUNITY	ENVIRONMENT	
Shooting and fishing	Tourism and recreation (general)		Landscape	Nature conservation
− Overgazing of moors − New farming methods may change habitat	+ Provides accommodation + Recreational possibilities − Restricted access	+ Employment + Some accommodation provided + Gives the area its basic character	Often maintains traditional landscape − Certain modern buildings detract from landscape	± Depending on type of cultivation or management
− Competes for grouse moors + Can be combined with pheasant rearing + Can provide good cover for deer	+ Provides recreational opportunities + Greater capacity to absorb visitors	+ Employment	+ Offers the possibility of concealing caravan sites . ± Landscape effect	± Valuable habitats may be gained, maintained, spoilt or lost
	− Quarry sites and traffic discourage visitors + Disused sites provide picnic places and recreational opportunity + Geological studies	+ Employment − Traffic	− Dust and noise − Traffic − Visual effect	± Important wild life habitats may be destroyed but disused working may become valuable later
± Affect angling	+ Water recreation facilities − Restriction of some access	+ Employment	± Landscape	± Development of new habitats may or may not not compensate for the loss of others
− Pollution of streams can reduce fish population	+ Provides some goods for the tourist market	+ Employment		− Pollution can affect fish and wildlife
	− Restricted access on certain days detrimental to visitors − Restricts boating etc.		+ Management of traditional gouse moors maintains the landscape	− Illegal shooting
+ Increased demand from tourists		+ Employment ± Effect on social character of community	− Some caravans and camping sites − Traffic + Economic use of old buildings for accommodation	
+ increased demand − Boating and swimming affect fishing − Fire risk on grouse moors	− Certain activities especially water conflict with each other e.g. swimming and canoeing − Too many visitors detract each others' pleasure	± Effect on social character of community	− Traffic on certain days − Litter − Overuse of certain places e.g. Malham	− Too many visitors can harm wildlife
	+ Accommodation for tourists			
	− Restrict caravan and camping sites + Need for conservation may prevent overuse	± Planning controls		+ Maintenance of flora and fauna
+ Enhances habitats	+ Increases tourist and educational potential	+ Enhances quality of life	+ Maintenance of diverse habitats	

Figure 10.4 Opportunities and agencies for change in a moorland ecosystem

		Opportunity	Main Statutory Bodies*
Single Agency Opportunities	Agriculture:	Promote farm amalgamation, reduce fragmentation	MAFF
		More intensive use of land	MAFF
		Reconsider marketing co-operatives	MAFF
	Quarrying:	Restrictions on heavy lorry routes	Local Authorities
	Tourism:	Survey of resources available	Tourist Boards
	Recreation:	Encourage time–space zoning of water	Regional Water Authorities
	Game management:	Intensification of managed field sports activities	No statutory body required
		Encourage farming community to integrate shooting with other activities	No statutory body required
	Community:	Traffic management	Local Authorities/National Park Authorities
		Maintaining level of services, such as education, sewerage etc.	Various appropriate authorities including
		Community participation	Local Authorities, Regional Water Authorities
			Local Authorities
		Housing policies	Local Authorities
	Landscape:	Conservation of Pennine towns	Local Authorities
	Nature Conservation:	Promote wild life conservation	Nature Conservancy Council
		Improve water-related resources	Regional Water Authorities
Dual Agency Opportunities	Agriculture:	Reconsider Common Land problem	MAFF, LAs/NPAs
	Quarrying:	National planning of mineral extraction	DI, LAs (vein minerals); DOE, LAs (limestone and aggregates)
		Develop rail connections	LAs/NPAs, British Railways Board
	Manufacturing:	Promotion of craft image	LAs, COSIRA
	Tourism:	Financial assistance	COSIRA, English Tourist Board
	Nature conservation:	More nature reserves	Nature Conservancy Council, LAs
		Educational uses	Nature Conservancy Council, LAs

Multiple Agency Opportunities

Category	Opportunity	Agencies
		British Waterways Board, RWAs, MAFF, Forestry Commission, LAs/NPAs
	Encourage diversification of sources of farm income	MAFF, Tourist Boards, FC
	Carry out land capability studies for agriculture and forestry purposes	MAFF, FC, LAs/NPAs, RWAs
Forestry:	Develop recreational and tourist potential of existing and new forests	FC, Tourist Boards, LAs/NPAs
	Carry out amenity tree planting	FC, LAs/NPAs, RWAs
	Integrate forestry and other land uses, such as shelter belts	FC, MAFF, NPAs, RWAs
Water:	Develop recreational potential of water	RWAs, BWB, Regional Sports Council, Tourist Boards, LAs/NPAs
	Diversify uses of catchment areas	RWAs, BWB, MAFF, FC, LAs/NPAs
Manufacturing:	Provide sites and/or buildings	LAs, COSIRA, DI
	Develop growth points	LAs, COSIRA, DI, Development Commission
Tourism:	Develop accommodation centres	LAs, Tourist Boards, Development Commission, COSIRA
	Improve tourist transport facilities	LAs/NPAs, transport operators
	Advice for farmers providing tourist accommodation	MAFF, Tourist Boards, COSIRA, Countryside Commission
	Information centres	LAs/NPAs, Tourist Boards, FC, RWAs, Countryside Commission
Recreation:	Develop more sites including a major recreational facility	LAs/NPAs, RWAs, BWB, FC, Tourist Boards Regional Sports Council
	Re-examine Hellifield proposal	RWAs, MAFF, LAs
	Access (and management) agreements	Local Authorities/National Park Authorities, Regional Water Authorities, FC
	Wardening, sign-posting, way-marking, creating footpaths	LAs/NPAs, RWAs, FC, CC
Community:	Encourage small-scale population expansion	LAs, COSIRA, DI
	Introduce more varied employment	LAs, COSIRA, DI
	Improved transport system	LAs/NPAs
Landscape:	Introduce comprehensive land-use planning	MAFF, LAs/NPAs, FC, RWAs, Tourist Boards
Nature Conservation:	Cooperation of other bodies	MAFF, LAs/NPAs, FC, RWAs, Tourist Boards

* The appropriate tier of local authority is not indicated. It should also be noted that the Peak Park Joint Planning Board combines the functions of Local Planning Authority and National Park Authority.

10.3 WATER RESOURCE SYSTEMS

10.3.1 Introduction

We have used a water resource system as an example at a number of previous places in the book. A very broad overall view was presented in Section 1.4.3; we outlined a spatial interaction model of water flows in Section 4.7; we presented some accounting equations, related to pollutants, in Sections 5.2.4 and 5.4.7; and we made some points on the network (and hierarchical) structure of river systems in Section 7.2.9. In this section, we draw the threads together and take a more detailed look at water resource systems in the light of the concepts and techniques now available to us.

In Section 10.3.2, we examine the basic description of a water resource system in relation to the basic concepts of systems analysis. Then, in the following three subsections, we present examples of models and their use in planning and control. A number of concluding comments are made in Section 10.3.6.

10.3.2 Description and basic concepts

We noted in Section 1.4.3 that the main natural process driving the system was the well-known hydrological cycle, and in Section 7.2.9 we made some brief comments on the evolution of river systems on minimum work principles. Here, however, we will concentrate on the construction of facilities by human agents for a variety of purposes and the associated modelling, control, and planning problems.

The main purpose of water resource systems is the supply of water, of adequate quality and at suitable pressures, for a variety of uses which can be broadly classified as agricultural, industrial, and domestic. The system has some subsidiary roles in some cases: hydro-electric power supply, flood control, and recreational uses are perhaps the most important.

The system, formally, can be seen as a set of nodes and links. The latter elements carry the flow and may be natural or manufactured. The nodes can be considered as sources and sinks, some of which may involve storage and control or other processes. The processes at a node could include water treatment, for example. A variety of components of such networks is shown in the different parts of Figure 10.5.

The sources are supply points. These may consist of the steady flow past a point of a river, some of which would be available for extraction later, or reservoirs, natural groundwater stores, or the product of a treatment plant involving re-use of water. The sinks can usually be treated as nodes representing the point at which water is extracted near demand areas. However, the design of associated pipeline networks is of considerable importance (as in the case of New York, described by de Neufville, 1974).

There are essentially two kinds of feature of the design problem: the size and location of major facilities such as reservoirs and distribution networks—the major 'capital' projects, and the design of an operating policy for a given system.

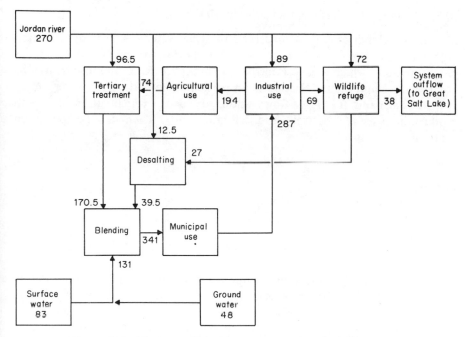

Figure 10.5 Elements of a water resource system, including re-use

As an overall optimization problem, the two features could be combined, but in the examples to be presented below we consider them separately. A particular feature of the operating policy is pricing, and since this involves a variety of users and impacts, this is of itelf one aspect of an overall economic analysis.

The variety of impacts demands a full cost–benefit analysis. However, this is very difficult in practice, not least because there are so many intangibles—such as recreational or flood control benefits. Further, the direct economic impact of water resource schemes is often very difficult to assess as water is only one factor input in the various sectors. More usually, therefore cost-effectiveness of cost minimization approaches are used.

Given this background, what is the systems analytical approach? The 'analysis' stage demands the availability of simulation models which represent the flow of water through the system for a stated distribution of facilities and operating policies. This of itself is a complex problem because of the high degree of interdependency, but perhaps most of all because of the stochastic nature of the system: the flows at any time are a complex function of the previous rainfall history (and system use).

The design task is a large-scale combinatorial problem. As we saw in Chapter 6, it is possible in principle to use optimization methods, and we will present some examples below. However, it is also possible to make some progress by normal manipulation of facilities in conjuction with a simulation model. De Neufville (1974), for example, suggests that it had been possible to 'save' $100 m. on a major scheme proposed for distribution in New York city by these means.

10.3.3 Example 1: a basic optimization model

Bishop and Hendricks (1974) observed that the water resource optimization problem has similarities with the transportation problem of linear programming — which was described above in Section 6.2.2. Suppose we use the notation of Chapter 4, and Section 4.7 in particular, and take O_i as water supply at i, D_j as demand at j, T_{ij} as flows, and c_{ij} as costs — which can include both production costs at i as well as distribution costs. Then the transportation problem is

$$\text{Min } Z = \sum_{ij} T_{ij} c_{ij} \qquad (10.13)$$
$$\{T_{ij}\}$$

such that

$$\sum_{j} T_{ij} = O_i \qquad (10.14)$$

$$\sum_{i} T_{ij} = D_j. \qquad (10.15)$$

Bishop and Hendricks show how to augment this problem to deal with water re-use. A demand point can also be seen as an origin point at which there would be a treatment cost. A part of such a network is illustrated in two ways — where the re-use is in-house, and where not — in Figure 10.6. Additional flow constraints must also be incorporated so that the flow out of a treatment plant does not exceed the inflow.

The other relaxation which is needed is to treat the equality constraints (10.14) and (10.15) as inequalities. This means that, in particular, the level of usage — or even construction size of a facility — can be determined in the programme, with O_i being an upper bound rather than a pre-determined exact value. The problem then becomes.

$$\text{Min } Z = \sum_{ij} T_{ij} c_{ij} \qquad (10.16)$$
$$\{T_{ij}\}$$

such that

$$\sum_{j} T_{ij} < O_i \qquad (10.17)$$

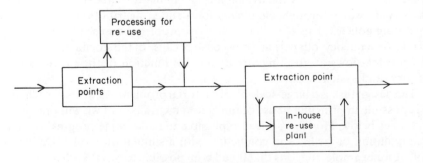

Figure 10.6 Two kinds of water re-use

$$\sum_i T_{ij} > D_j \qquad (10.18)$$

$$\sum_j x_{lj} < \sum_i x_{il} \qquad (10.19)$$

with the last equation applying to any node l at which there is a treatment plant. The quantity P_i, say, given by

$$P_i = \sum_j T_{ij} \qquad (10.20)$$

then gives the level of supply from i. This will be influenced by the prescribed upper band O_i, but more by the production and distribution costs at the origin. This last aspect is perhaps made more explicit if we replace c_{ij} by $C_i + c_{ij}$, where C_i represents the production cost at i and c_{ij} the distribution cost to j. The objective function could then be written

$$Z = \sum_{ij} T_{ij}(C_i + c_{ij}). \qquad (10.21)$$

There are two possible inadequacies with this formulation. First, the costs C_i and c_{ij} are likely to be nonlinear functions of total production at l, P_i, and flow from i to j, T_{ij}. Secondly, it may not be easy to assign precise proportions of water from particular origins to particular destinations unless the river and pipeline network happens to be very sparse. We now show how to tackle both of these problems and thereby illustrate the use of nonlinear programming (c.f. Section 6.2.3) and the water flow model proposed in Section 4.7.

We can hypothesize that C_i and c_{ij} are functions of P_i and T_{ij} and take the kind of forms shown in Figure 10.7. This indicates economics of scale up to a certain point and diseconomies after that. The form of the functions in the figure imply that there is an optimum size of reservoir at a particular location, or that there is an optimum pipeline size for a certain flow between locations. (For the present, we neglect the 'assignment' nature of the network problem, though return to it later.) However, it does not, of course, follow that for a system each locational unit, or each pipeline unit, is at its individual optimum size. The fact that it is not can, indeed, be considered to be a systemic effect.

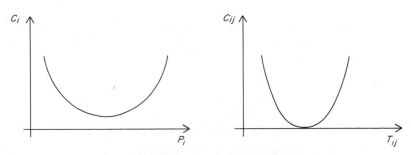

Figure 10.7 Water: storage and transport costs

Since the heart of the problem of determining facility sizes is now in the cost nonlinearities, we can revert to using equality constraints. We write the costs as functions of usage to make the nonlinearity explicit. The problem can then be stated as

$$\text{Min}_{\{P_i, T_{ij}\}} Z = \sum_i P_i C_i(P_i) + \sum_{ij} T_{ij} c_{ij}(T_{ij}) \tag{10.22}$$

such that

$$\sum_j T_{ij} = P_i \tag{10.23}$$

$$\sum_i T_{ij} = D_j \tag{10.24}$$

together with any other constraints relating to re-use or whatever, and the usual non-negativity constraints. This problem was first set out by Wilson (1973b) and we now mainly follow the subsequent argument of that paper, though with some additions.

First, as a preliminary, note that the destination label j could be a subscript list and could refer to both sector and location. In principle, the water authority could charge, say p_j per unit for sector-location j. Then if the demand function for water was known in each sector, say written as $D_j(p_j)$, net revenue or profit would be

$$\Pi = \sum_j D_j(p_j) p_j - \sum_i P_i C_i - \sum_{ij} T_{ij} c_{ij} \tag{10.25}$$

and the problem could be reformulated to maximize Π subject to the previous set of constraints.

Secondly, as a further preliminary comment, note that the cost functions in Figure 10.7, are similar to those introduced in Section 8.2.2 in a discussion of the evolution of trunk routes in networks. The same kind of broad consideration is likely to apply here: it may be appropriate to complicate the constraint set in such a way that the kind of programme presented above can be combined with the trunk-link-evolution ideas of Section 8.2.2 so that we can model the evolution of a water network with a hierarchical structure.

We now consider two developments of the nonlinear problem given by (10.22)–(10.24). The first is a linear approximation, and the second involves the spatial interaction model of Section 4.7 and ultimately leads to an alternative linear approximation.

If the form of the curves in Figure 10.7 were known, and if the area around the minimum was fairly flat in each case, then as an approximation it may be possible to consider C_i and c_{ij} each to be fixed within ranges:

$$P_i^{\min} < P_i < P_i^{\max} \tag{10.26}$$

$$T_{ij}^{\min} < T_{ij} < T_{ij}^{\max} \tag{10.27}$$

The problem then becomes, approximately, the following linear problem:

$$\text{Min}_{\{P_i, T_{ij}\}} Z = \sum_i P_i C_i + \sum_{ij} T_{ij} c_{ij} \tag{10.28}$$

such that

$$\sum_j T_{ij} = P_i \qquad (10.29)$$

$$\sum_i T_{ij} = D_j \qquad (10.30)$$

$$P_i < P_i^{\text{max}} \qquad (10.31)$$

$$P_i > P_i^{\text{min}} \qquad (10.32)$$

$$T_{ij} < T_{ij}^{\text{max}} \qquad (10.33)$$

$$T_{ij} > T_{ij}^{\text{min}} \qquad (10.34)$$

and, of course, the lower 'minimum' levels could be taken as zero. It may also be possible to do model runs of this kind and then to modify the approximations built into the constraints (10.31)–(10.34) in the light of that experience.

Another way of dealing with nonlinearity would be either to make piecewise-linear approximations for the cost functions, or to assume step functions and have fixed costs within 'ranges'. This second idea is a more elaborate version of the problem given by (10.28)–(10.34). Let us take the second technique as an example and apply it to pipelines only. Let k denote categories of pipeline size and assume costs are independent of flow within a category and can be taken as c_{ij}^k. Define a new variable x_{ij}^k which is to be 1 if a k-pipeline exists between i and j, 0 otherwise. Then the pipeline cost term becomes $\sum T_{ij} x_{ij}^k c_{ij}^k$, and there is an additional constraint

$$\sum_k x_{ij}^k < 1 \qquad (10.35)$$

implying that at most one pipeline can connect i and j.

Finally, let us return to the problem of the distribution variable, T_{ij}. We saw in Section 4.7 that, in a complex water network, such flows might be modelled on a spatial interaction basis as

$$T_{ij} = A_i B_j O_i D_j e^{-\beta c_{ij}}. \qquad (10.36)$$

Suppose the flow aspects of this are 'relatively constant' and that facility size at the origin is the important planning variable. Define, therefore,

$$g_{ij} = A_i B_j D_j e^{-\beta c_{ij}} \qquad (10.37)$$

so that

$$T_{ij} = O_j g_{ij} \qquad (10.38)$$

and assume that the $\{g_{ij}\}$ can be treated, as an approximation, as a set of constant coefficients. Then our earlier, basic, problem becomes

$$\underset{\{O_i\}}{\text{Min}} \ Z = \sum_i O_i C_i + \sum_i O_i \sum_j g_{ij} \qquad (10.39)$$

subject only to

$$\sum_i O_i = \sum_j D_j = \text{constant} \tag{10.40}$$

since the detailed condition that demands at j are met has been built into (10.37) and (10.38) This can be written

$$\underset{\{O_i\}}{\text{Min}} Z = \sum_i O_i(C_i + \sum_j g_{ij}) \tag{10.41}$$

subject to (10.40), showing that g_j is being treated as a quasi-cost.

With the g_{ij}'s as constant, this is a well-known linear programming problem known as the 'knapsack' prblem, and can be solved by standard methods.

10.3.4 Example 2: reservoir discharge policies

In the first example, we considered the optimum locations and sizes of major capital facilities like reservoirs and pipelines. Here, we use a problem formulated and analysed by Leclerc and Marks (1974) to illustrate the second kind of planning and control problem: the optimal management of an existing system. This particular problem is concerned with the optimum discharge policy for a system of reservoirs. It is the interconnectedness of the elements of the system which makes the solution to this problem non-obvious, and optimization methods can be used.

Here, we consider a slightly simplified view of Leclerc and Marks' problem, particularly in the notation used. They use the concept of reservoir drawdown, measured in height of water. We will use the concept of reservoir store only, so that when they minimize drawdown, we maximize storage. Because there is a linear relationship between the two, this does not affect the essential nature of the treatment of the problem. We also offer a more general formulation of the constraint equations. We consider first the programme as formulated by Leclerc and Marks and then some variants of it.

The Leclerc and Marks (hereafter, LM) programme is not easy to follow because they attempt to tackle a difficult problem and do not make it entirely clear in their notation how it is resolved. That is, the flows in the system are stochastic but the decision variables of their problem refer to 'seasons' which, for illustration, they take as months of a year. Thus, their time label, t, refers to months and their season label, i say, to a particular month. Thus i is the remainder of 't divided by 12'. That is, if $t = 1, i = 1$, if $t = 13, i = 1$, and so on. The constraints are first formulated in terms of t, and then (though it is not entirely clear) in terms of i.

The basis of the solution is the so-called linear operating rule

$$d_{jt} = S_{jt-1} - b_{ji} \tag{10.42}$$

where S_{jt-1} is the stock in reservoir j at the end of $t-1$, d_{jt} the discharge in period t, and b_{ji} is a 'seasonal' adjustment factor which is taken as the main decision variable. (Recall that i is, formally, a function of t as described and should be written as $i(t)$, to leave t and j as the only 'free' subscripts in equation (10.42)—but we will leave it as i with this understood.)

Let $r_{j,t}$ be the flow during period t into reservoir j. Then the stock at the end of t is clearly given by the accounting equation

$$S_{jt} = S_{jt-1} - d_{jt} + r_{jt} \qquad (10.43)$$

and, using (10.42), this can be written

$$S_{jt} = b_{ji} + r_{jt}. \qquad (10.44)$$

Note, from equation (10.42)—and (10.44)—that another way of describing b_{ji} is as the remaining stock after discharge and before inflow, but which is taken as fixed by season.

We now offer one development of the LM-notation. Consider Figure 10.8, which shows the system they use to illustrate their solution to the problem. Each streamflow is labelled according to the particular reservoir it goes into. Let R_j denote the set of streams which contribute to flow j and let D_j be the set of discharges which contribute to it. Thus, for the LM-example, all the R_j's are made up of one streamflow while D_4 is the set $\{1, 2, 3\}$ and D_9 is the set $\{4, 5, 6, 7, 8\}$.

The objective function used to determine optimum operating rules is to minimize reservoir drawback, or, equivalently, to maximize storage. Thus we can

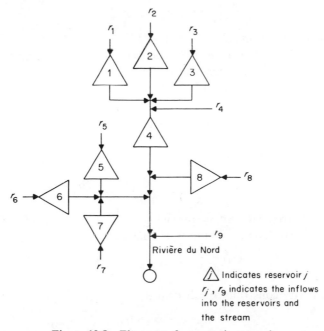

Figure 10.8   Elements of a reservoir network

write

$$\text{Max}\, Z = \sum_j C_{ji} S_{ji} \qquad (10.45)$$
$${\{b_{ji}\}}$$

where S_{ji} is *mean* storage in season i and C_{ji} is a cost associated with that level of storage at j. (This may be a loss incurred through fall from maximum because of loss of recreational use, for example.)

The constraints are stated in terms of t-subscripts, not i, for the present. There are five types. First, there is a continuity equation for each reservoir which can be written

$$S_{jt} = S_{jt-1} - d_{jt-1} + \sum_{k \in R_j} r_{kt} + \sum_{i \in D_j} d_{it}. \qquad (10.46)$$

The second constraint arises from a requirement for a minimum streamflow at the 'tail of the system' during each period, say $q_t^{\min}$. From the interpretation of b_{ji} above, we can see $b_{ji-1} - b_{ji}$ net addition to flows from the reservoir, and so if we add total exogenous inputs, we have

$$\sum_j (b_{ji-1} - b_{ji} + r_{jt}) > q_t^{\min}. \qquad (10.47)$$

The second type of constraint arises from discharges being non-negative:

$$d_{jt} = S_{jt-1} - b_{ji} > 0 \qquad (10.48)$$

and finally so that the storage capacity of each reservoir is satisfied:

$$S_j^{\min} < S_{jt} < S_j^{\max} \qquad (10.49)$$

using an obvious notation.

The problem is turned into a *period* to determine b_{ji} by replacing all t-indexed variables by an appropriate mean of a stochastic process at a risk level α. Thus r_{jt} is replaced by $r_{ji\alpha}$, and $S_{ji\alpha}$ is calculated rather than S_{jt}. The whole problem can then be stated as

$$\text{Max}\, Z = \sum_j C_{ji} S_{ji\alpha} \qquad (10.50)$$
$${\{b_{ji}\}}$$

subject to

$$S_{ji\alpha} = b_{ji} + \sum_{k \in R_j} r_{ki\alpha} + \sum_{i \in D_j} (S_{ii-1\alpha} - b_{ii}). \qquad (10.51)$$

This is derived from (10.46) by replacing d_{jt-1} by $S_{ji-1\alpha} - b_{ji}$ and d_{it} by $S_{ii-1\alpha} - b_{ii}$ similarly. The minimum flow constraint is

$$\sum_j (b_{ji-1} - b_{ji} + r_{ji\alpha}) > q_i^{\min}. \qquad (10.52)$$

The non-negativity constraint on discharge is

$$S_{ji-1\alpha} - b_{ji} > 0 \qquad (10.53)$$

and the reservoir capacity constraints are

$$S_j^{min} < S_{ji\alpha} < S_j^{max}. \tag{10.54}$$

This problem can be solved for a variety of values of α, expressing risk levels, and a set of $\{b_{ji}\}$ selected for a level of risk which is considered appropriate.

The optimization strategy used by Leclerc and Marks is a clever way of dealing with the stochastic nature of the system's water inputs. There may, however, be an alternative strategy: to work with the stocks and discharges directly. Suppose we keep the objective function form but write it in terms of time totals as

$$Z = \sum_{jt} C_{jt} S_{jt} \tag{10.55}$$

and rewrite the constraint equations so that d_{jt} replaces b_{jr}, and all r-indexed variables are replaced by t-indexed variables. We should also note that the cost C_{jt} in the objective function could include a discount term so that the near future was given greater weight than the more distant future, which would be sensible given the uncertainties.

The first set of constraints are now the accounting equation (10.46), repeated here for convenience:

$$S_{jt} = S_{jt-1} - d_{jt-1} + \sum_{k \in R_j} r_{kt} + \sum_{i \in D_j} d_{it}. \tag{10.56}$$

The minimum flow constraints are generated from (10.47) by replacing the b_{ji} terms using (10.44). This leads, after some manipulation, to

$$\sum_j (d_{jt} - r_{jt-1} + r_{jt}) > q_t^{min}. \tag{10.57}$$

The non-negativity constraints can be written, using (10.43):

$$S_{jt-1} - S_{jt} + r_{jt} > 0. \tag{10.58}$$

Finally, the range of possible values of reservoir storages are as before

$$S_j^{min} < S_{jt} < S_j^{max}. \tag{10.59}$$

The problem now becomes

$$\underset{\{S_{jt} d_{jt}\}}{\text{Max}\, Z} = \sum_{jt} C_{jt} S_{jt} \tag{10.60}$$

subject to constraints (10.56)–(10.59).

There are two difficulties which have to be noted immediately with this formulation. First, there is a large number of variables because of the need to range over a large number of time periods as well as sectors. Secondly, we have not specified how to deal with the stochastic variables r_{jt}. The two problems are related. One can imagine a smooth 'steady state' mode of behaviour for the system under 'normal' conditions, but different discharge policies being needed for different kinds of drought. The number of time periods which it is necessary to incorporate into the analysis will be determined by the relaxation time: the time needed to get over transient effects after a perturbation like a drought. This problem could be investigated by simulation with the model using sets of $\{r_{jt}\}$ data which represented a variety of conditions which had been observed in the past. This would help determine discharge policies for a variety of initial conditions and the results could then be assessed for a range of possible $\{r_{jt}\}$ sets for the future.

A final problem was noted by Leclerc and Marks in connection with their formulation and can be extended to the alternative formulation above. For the LM-strategy, no account is taken of the balance of withdrawal across the reservoirs: it would be possible to empty a small number of reservoirs altogether and leave the rest almost untouched. In the second formulation of the problem above, this could also arise, and in addition the same effect could happen over time: there could be large withdrawals in only one or two time periods. Leclerc and Marks propose tackling this problem by the use of a mini-max criteria. This involves replacing the objective function by one which minimizes the drawdown from the reservoir which had the largest weighted drawdown:

$$\operatorname*{Min}_{\{G_{ji}\}} Z = \left[\operatorname*{Max}_{\{G_{ji}\}} C'_{ji} G_{ji} \right] \tag{10.61}$$

where G_{ji} is drawdown at j in period i and C'_{ji} the unit cost of drawdown at j in i. In storage terms, this would involve a maxi-min criterion. It is perhaps simpler to proceed with additional constraints—for example that no storage loss is greater than $x\%$ of total loss in the period:

$$S_{jt} - S_{jt-1} < x \sum_j [S_{jt} - S_{jt-1}]. \tag{10.62}$$

By experimenting with various values of x, this may lead to suitable solution.

10.3.5 Example 3: a note on optimization and pollution control

In Section 5.4.7, we outlined model equations due to Kelly and Spofford (1977) which predicted the concentration of pollutants as well as biological populations in the reaches of a river. They argue that these can form the basis of optimization methods for pollution control. However, they only sketch the corresponding mathematical programmes, but they offer useful insights on a process which may become feasible in the future.

Let $\{W_i\}$ be activity levels of processes which are used to reduce the emission of pollutants $\{Z_i\}$. Let $\{c_i\}$ be the unit costs associated with the reduction processes. Then Kelly and Spofford suggest a linear programme of the following kind:

$$\operatorname*{Min}_{\{W_i\}} C = \sum_i c_i W_i \tag{10.63}$$

subject to

$$\sum_i (A_{1i} W_i + A_{2i} Z_i) > B_i \tag{10.64}$$

$$\sum_i H_i Z_i = X_i \qquad (10.65)$$

$$X_i \begin{cases} < \\ > \end{cases} S_i \qquad (10.66)$$

and non-negativity constraints

$$Z_i > 0, \quad W_i > 0, \quad X_i > 0. \qquad (10.67)$$

The objective function, (10.63), is self-explanatory. The constraints (10.64), involving coefficients $\{A_{1i}\}$ and $\{A_{2i}\}$ and quantities $\{B_i\}$, guarantee that enough of the required set of products will be generated by industry. This is expressed in terms of a mix of expenditure of waste-reduction ($\{W_i\}$) and 'permitted' emissions ($\{Z_i\}$). Equations (10.65) are a linear form of the accounting equations (5.91). The coefficients $\{H_i\}$ convert the Z_i's to pollutant concentrations, $\{X_i\}$. The constraints (10.66) then form the standards which are to be satisfied.

This is a well-defined programme if all the relationships involved are linear or can be treated as such in an approximation. They point out, however, that their own model (5.91) is nonlinear. They indicate how, when (10.65) is replaced by a set of nonlinear equations, it is then more convenient in practice to treat the constraints as penalty functions which are added to the objective function. The problem then becomes

$$\min_{\{W_i\}} F = \sum_i (c_i W_i + p_i) \qquad (10.68)$$

subject to

$$\sum_i (A_{1i} W_i + A_{2i} Z_i) > B_i \qquad (10.69)$$

and

$$Z_i > 0, \quad W_i > 0 \qquad (10.70)$$

where p_i is the penalty associated with

$$X_i = h_i(Z_i) \qquad (10.71)$$

exceeding the standard S_i. Thus the nonlinear 'biological' relations (10.71) have to be supplied together with a (possibly nonlinear also) penalty function. This also, therefore, gives us an insight into a new technique for nonlinear programming problems.

10.3.6 Concluding comments

The scale of complexity of water resource systems is an interesting one for systems analysis. It is certainly sufficiently complicated for analytical techniques to be useful, but, for once, sufficiently simple for optimization methods to be usable as part of the design and planning process. However, it could be argued that this is because we have not treated the hydrological variables at a sufficient level of detail. This may represent one of the directions of future research: to find new ways of dealing with detail and particularly the stochastic nature of the input variables. Otherwise what is needed is much more experience with the kinds of models described — and some of greater complexity have been used and are cited in the bibliographic section below — together with an extension of the evaluation side away from cost minimization towards cost-effectiveness and cost–benefit analysis. Relatively little progress in this direction seems to have been made in spite of the fact that the first model-based studies (Maass, *et al.*, 1962) were rooted in cost–benefit analysis.

10.4 URBAN SYSTEMS

10.4.1 Introduction

We concluded at the end of the last section that water resource systems were probably at the level of complexity at which systems analytical methods were at their best. There is an obvious sense in which urban systems, seen as whole systems, are beyond this. In previous chapters, however, we have seen many examples of systems analytical methods applied to urban *sub*systems and this, of itself, is valuable. Paradoxically, the fact that the methods cannot effectively deal with the complexities of the whole system emphasizes the importance of ongoing research: the problems to be tackled are urgent and systems research could yet make a more effective contribution.

In this section, we follow, broadly, the organization of the previous two. In Section 10.4.2, we review the description of urban systems and associated basic concepts. Then, rather than focusing on a range of particular examples — partly because many have already been presented and partly because the field is so huge — we concentrate on reviews of the state-of-the-art followed by a single example. We first do this in relation to analysis in general terms by studying model building in Section 10.4.3, with the outline of the example following in Section 10.4.4. We then consider the use of models in planning in Section 10.4.5. A number of concluding comments are made in Section 10.4.6.

10.4.2 Description and basic concepts

In Section 1.4.4, we presented a very brief sketch of the city as a system — identifying the main subsystems and some of the problems associated with them. In this subsection, we examine the elements of a city and their interrelationships in more detail as a preliminary to reviewing progress in modelling and planning. We can begin by distinguishing the following kinds of elements:

(i) people;
(ii) organizations;
(iii) commodities;
(iv) physical structures and associated facilities: buildings, transport systems, etc. (*p*-structures);

(v) activities of people and organizations;

(vi) aggregate entities, like neighbourhood properties;

(vii) relations between elements (r-structures).

Each group of elements has to be further subdivided into categories which are useful to increase our understanding of urban change, and as an aid to planning.

Consider, for example, some of the following ways of characterizing the elements defined above

(i) people, by
   —sex
   —age
   —position in household
   —education history
   —job and workplace location
   —social class
   —income
   —wealth
   —car availability
   —house type and residential location
   —'basket' of goods bought and associated locational patterns
   —'basket' of services used and associated locational patterns (schools, hospitals, banks, ...)
   —recreation pattern
   —allocation of time budget
   —range of choice of activities
   —r-structures with other people and organizations.

(ii) organization, by
   —role or purpose (manufacturing, service supply, social, governmental, etc.)
   —number of employees of different types
   —buildings used, and their location(s)
   —economic structure (capital, cash flows, etc.)
   —production structure (commodity inputs, product or service outputs)
   —r-structures with other people or organizations.

(iii) commodities
   —identification (chemical, energy, etc.)
   —possible roles
   —structure in production or service processes
   —use patterns: consumption and production
   —price
   —'market' organization.

(iv) p-structures
   —physical description (buildings, associated equipment, transport networks and vehicles)
   —land use and location

—people or organization: (a) owning
(b) using
—associated economies (value, rent, etc.)
—role; possible alternative roles; production function.

(v) activities
—description (people, organization, p-structures and r-structures operating in combination—a behaviour pattern of a subsystem or system)
—role or purpose
—place in wider structure
—inputs
—outputs
—role in r-structure.

(vi) aggregate entities
—e.g. neighbourhood environment, by
—p-quality (housing quality, environment, etc.)
—number of residents
—number of organizations by type
—social class of residents
—'noxiousness' of organizations
—density of residents
—density of workers
—level of traffic congestion.

(vii) r-structures
—identification and description: links between individuals and organizations to carry out activities.

The schema sketched above implies the existence of a large number of 'overlapping', interacting, subsystems. These can be defined in relation to most of the major types of element: people and organizations are obviously well-defined systems; so are buildings, certain kinds of activities, and so on. There are also a number of 'resolution' issues in subsystem division: individual people and organizations are subsystems, but so also are groups of these, such as the members of a social class, the residents of a neighbourhood, the employees of a firm, or the set of organizations which make up a 'sector', like the 'motor industry'. Ideally, the analyst would like to have data and information available at the final level of resolution so that it is possible to aggregate in a variety of alternative ways for different purposes. In practice, it is impossible to achieve this. In particular, data sources are partial (and themselves based on different resolution levels—like different spatial units). This means, for example, that there will be no data which list all the characteristics together—for people, say, as in the list above—which is one reason for the development of methods such as micro-simulation, as we saw in Chapter 8.

The main processes associated with people are concerned with demography and subsistence. Population changes through birth, death, and migration and people need a minimum of resource inputs to live: shelter, food, education and health services, and so on. These resource inputs are normally achieved through work or through some sort of social support which is, or arises out of, the work of others. The lifestyle of people is therefore closely linked with associated economic and social systems which provide goods and services and the opportunity to work. Economic development can be seen as the product of technological change and the evolution of organizational principles, such as the division of labour, which facilitate growth. Social development can be seen as arising from a variety of mechanisms according to viewpoint. It may arise out of a concern for social welfare from the better-off to the less well-off; or as the provision by a capital-supporting state to facilitate the reproduction of labour power and the maintenance of a reserve pool of labour; or as the result of class struggle, with the less-well-off forcing concessions from the better-off with the state acting as a mediator. It may be a mixture of all these mechanisms, and more. Buildings decay and become obsolescent; new ones are constructed. The relative importance and availability of commodities changes. Relational structures evolve and change.

Within this broad backcloth, there is an immense variety of motivations which generate a corresponding wide variety of processes. And it is against this backcloth that we must assess the nature of urban change and its current problems. Cities can be seen as arising from different aspects of economic development: the need for a place to market agricultural goods; as centres of trade and diplomacy; as high-density labour forces for large factories and other organizations; and as social centres. The development of western cities over the last two hundred years can then be charted as the product of economic growth coupled with processes of decentralization and social differentiation, the latter two aspects being closely linked.

The spatial pattern and dynamics of urban change are determined by these processes and their interdependencies. The complexity of the interactions between subsystems generates a wide range of systemic effects. One of the main sources of the latter are what we called aggregate entities in the list of elements. People, for example, will decide where to live by considering the characteristics of neighbourhoods in relation to what they can afford. These characteristics are the aggregation of a lot of individual decisions. Firms will want to locate in relation to other firms with whom they have dealings and this often leads to clustering of particular sectors — like the financial sector of many cities for example. But these processes can also lead to rapid change. If the social character of a neighbourhood begins to change — for example through a large influx of poorer immigrants at a time or urban growth — then the richer residents will seek new territory, probably only available on the edge of the existing city or in surrounding villages, and the growth process will reinforce social differentiation.

It does not follow that the working out of these processes will always have good results. At present, for example, consider the consequences of the linked problems of energy shortages and corresponding price rises, rising unemployment, and inner city decay. Oil price rises (and possibly other factors) have provoked economic recession. This has led to increased unemployment and this has a differentially hard impact on the less-skilled. This leads to a greater demand for social services at a time when, because of the recession, there is increasing pressure on public expenditure. Both public and private sources of investment for the improvement of inner city areas decline. The problem is further complicated by the fact that the spatial structure of the city is determined by high levels of car-ownership among non-inner-city residents and there may be an impact through this as oil prices continue to rise and a critical point may be reached at which the underlying structure will begin to change.

In the next subsection, we review the history of models which help us to understand these processes and include a particular example. In the following two subsections, we present a particular example and the explore the role of models in planning and their use in policy formulation relations to the kinds of problems sketched above.

10.4.3 Urban models: a summary of the state-of-the-art

In Figure 1.9, we set out the main urban subsystems; this is repeated here for convenience as Figure 10.9 as it provides a useful framework for a description of the state-of-the-art in urban modelling. As a preliminary step, it is useful to interpret the main subsystems identified on the figure in terms of the description of the elements of the systems in the previous subsection. Broadly speaking, the left-hand side of the diagram is concerned with people and their activities, the right-hand side with organizations and their activities. These are each seen as generating physical structure and transport structure and flows. There are strong interdependencies between the subsystems. For example, the facilities (houses, shops, schools, etc.) used by people in their activities are supplied through organizations, which also supply jobs. These interdependencies are explicit in mathematical models of the subsystems through common variables. If a subsystem is modelled on its own, then variables which are in fact determined in another subsystem are treated as exogenous.

We have used models of the various subsystems to illustrate the various modelling methods which were introduced in Chapters 4–8. Here, therefore, we will simply recap on these in the context of a brief historical survey and review of the position which has been reached and then, in the next subsection, we present one integrated model as an illustration which illustrates this discussion.

Figure 10.9 also includes an implicit breakdown by spatial scale. Demographic and economic models on the left- and right-hand sides of the diagram are usually constructed for the whole study area. The models associated with the activity subsystems are then used to break the spatially-aggregated totals down to subareas within the city. In each case, the models are based on accounting principles; examples were given in Chapter 5. They each have histories which go back to the 1940's (Leslie, 1945, Leontief, 1951) but were only applied to urban systems on any substantial scale from the 1960's onwards (probably because, as with other urban models, they needed large computers). Their results can be presented in account form and examples of each, for West Yorkshire, can be found in Wilson, Rees, and Leigh (1977). As can be seen, and is clear from the

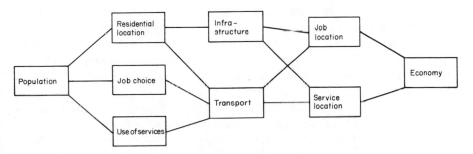

Figure 10.9 The main components of an urban system (Wilson, 1972)

definition of, and discussion of, the associated variables in Chapter 5, a tremendous amount of detail is involved. There can then be a problem of integrating this with other submodels, but we will defer the associated discussion until Section 10.4.6.

There is one problem, for any piece of analysis for a single city or region, which should be mentioned here which was only touched on in Chapter 5: any such system is part of a wider inter-urban or inter-regional system, because, of course, the systems are open. In the demographic case, this means that further submodels of migration flows have to be added, and, in the economic case, submodels of base flows are required. These can be constructed, based on spatial interaction principles, but there are often difficulties in practice because of paucity of data — especially in the economic case.

The population activity models are concerned with study areas divided into spatial zones. These are assumed to contain (generated from other 'economic' submodels) houses of different types, jobs (which is how individuals, in this sense, perceive the spatial distribution of economic activity), and a variety of service facilities. The models show how the individuals and households of the city are allocated to these different kinds of facilities. Since the various facilities, in relation to a household, say, are located in different places, this also generates the person transport pattern: the journey to work, shop, school, and so on. It is this feature which has led to many of the models being based on spatial interaction principles. The decision on choice of house, for example, is assumed to be related to a prior choice of workplace and may also be influenced by access to shops, schools, and other facilities in the neighbourhood of any housing which is being considered.

The basic principles for constructing models of this type are outlined in Sections 4.2 and 4.3. They can be derived by a variety of means and have a long history. In Chapter 4, we used entropy maximizing methods, though we demonstrated the alternative random utility methods in Section 6.2.6. These models have again been in common use since the early 1960's. (In fact, the reason for the upsurge at that time was almost certainly due to the development of large computers then rather than the intellectual difficulty of the problems at earlier times.)

The random utility approach demonstrates how spatial interaction models can be developed using the economic theory of consumers' behaviour: that preferences can be expressed in terms of utility functions and that each household maximizes this subject to any constraints. The addition of a random element, to represent market imperfections such as inadequate knowledge of available opportunities, or differences in preferences which have not been captured due to aggregation (or the analyst's inadequate knowledge), are essential to the generation of models in 'standard spatial interaction form' (though there are some important differences from other procedures). This feature has also been tackled from another direction. Economists developed models without the random element—for example in linear programming form as in the Herbert–Stevens model of residential location. It was shown in the early 1970's, however, that there is a close and hitherto unsuspected relationship between entropy maximizing methods and such economic models: the former can be seen as an 'imperfect market' version of the latter, and hence may be more realistic. Further, the kinds of assumptions which had been incorporated only in the economic models can now be added to entropy maximizing models so that they are no longer completely dependent on the old spatial interaction principles. This leads to great flexibility in model design.

The models discussed above for the spatial distribution of population activities all rely on physical, structural, and organizational variables as exogenous inputs. These are variables like H_i^k, the number of type k houses in each zone i, or W_j, a measure of the 'size' of shopping facilities in zone j. Such variables between them characterize the spatial–physical structure of the city and population and organizational activities can be seen as overlaid on these. We explained in Section 4.6 that this was not a serious problem in the use of models in planning because, usually, they could be taken as 'controllable' variables and manipulated as part of the planning process. However, this is not always the case. There are many circumstances where organizations develop according to their own goals subject to only the weakest kind of development control. And in any case, it is useful to see if we can find any economic–geographical explanation as to why urban settlement structures take the forms which they do. This, then, generates a new kind of modelling problem which, in the main, has only been explored relatively recently: how to model variables like $\{H_i^k\}$ and $\{W_j\}$. Some progress has now been made with this problem. It involves adding hypotheses about organizational behaviour. In the shopping centre case, for example, it can be assumed that entrepreneurs match shopping centre size (W_j) with revenue which can be obtained (D_j, say). The simplest hypothesis would be

$$D_j = kW_j \tag{10.72}$$

where D_j is calculated from the usual spatial interaction model of consumers' behaviour as

$$D_j = \sum_i S_{ij} \tag{10.73}$$

with

$$S_{ij} = \frac{e_i P_i W_j^{\alpha} e^{-\beta c_{ij}}}{\sum_k W_k^{\alpha} e^{-\beta c_{ik}}}. \qquad (10.74)$$

The additional equations (10.72) mean that the whole system, (10.72)–(10.74), can now be solved for $\{W_j\}$ as well as $\{S_{ij}\}$ and the stability of the equilibria thus obtained can be analysed.

In principle, this argument could be extended for other organizational or structural inputs to population activity models. It is likely to be possible for housing, but perhaps also for some of the variables associated with different kinds of firms. Much of this, however, is a matter for future research.

There is one obvious weakness with the argument about structural modelling presented above, especially with hypotheses like that implied by (10.72). That is, such hypotheses imply strong equilibrium assumptions: as consumer demand changes, it is assumed that there is a matching change in facility size. The dynamic evolution of physical and organizational structures will be much more complicated than this. There will be time lags, partly in response to entrepreneurial caution in new building, partly because of an inertial commitment to what is already there. This leads to a general concern with dynamic modelling, the principles of which were presented in Chapter 8. As we saw there, this field of work was initiated and stimulated by the publication in 1969 of Forrester's *Urban Dynamics*. Some progress is now being made with a variety of techniques, some of which we have already seen, but which will be illustrated in the example to be presented in the next subsection. This is geared towards a model-based account of the evolution of settlement and associated population and organization activities.

So far in the discussion above we have concentrated on subsystem models, though noting that integration is necessary. The main initiator in this sense was Lowry with his *A model of metropolis* in 1964, though in a sense Christaller and Losch's central place theories of the 1030's and 1940's can be seen as previous attempts from a different background. Lowry made a number of simplifying assumptions about some of the submodels involved. For the aggregate economic sector, he distinguished basic (exporting, not dependent on local markets) and nonbasic (essentially service) employment and assumed them to be related by a simple mechanism—that total employment could be generated from basic employment by a simple multiplier. The demographic model then consists of a simple assumption that the total population is proportional to total employment.

The heart of Lowry's model is a pair of interdependent spatial interaction models: residential location, which allocates people in relation to workplace; and service employment location, which allocates service jobs in relation to people. It is the feedback of people on jobs and vice versa which creates the interesting degree of interdependence as we will see in the illustration in the next subsection.

Lowry also has some interesting mechanisms in relation to land, but otherwise makes further simple assumptions about physical structure: that it appears to 'meet' demand (subject to a number of constraints). It is possible to add the type of structural-evolution mechanisms mentioned earlier to improve this part of the model, as we will also see in the illustration.

The original Lowry model was an important landmark in urban modelling, because it showed the importance of linking subsystem models. It could be improved with the addition of better demographic and economic models, and of interaction and structural models, but the discovery of the mutual feedback of population and service jobs in important and this feature persists in many models which do not otherwise look like Lowry's original.

We can now take this general discussion further in the light of a particular example, proceeding to a discussion of models as aids in planning in the following subsections.

10.4.4 An example of an urban model

We present, here, a Lowry-type model at a sufficiently fine level of resolution to show in principle how the various phenomena discussed in the previous two subsections can be incorporated. This means that a large number of subscripts and superscripts sometimes have to be used, but the heart of the model remains the two spatial interaction models of service use and residential location which have the same general characteristics as the simpler ones which were introduced to illustrate singly-constrained spatial interaction models in Section 4.3.

The overall structure, and sequence of calculation, of the model is shown in Figure 10.10. Initially, it is common to take only the spatial distribution of basic employment as given, but there is no reason why existing service employment should not be taken into account also. We present below the main model equations with only a brief commentary and then discuss its overall working. For convenience, the variables are defined together in Table 10.1.

Total employment in zone j in income group w is given by

$$E_j^w = \sum_g y^{wg} E_j^g + y^{wB} E_j^B. \tag{10.75}$$

The E_j^B's are given; employment by j in different service sectors g are generated in the service model to be described below. The y^{wg} and y^{wB} are coefficients which give the proportions of jobs in different sectors by income group. This is important because it is the main source of income differentiation in the model.

The residential location model is a disaggregated version of that presented as equations (4.39) and (4.40). Superscripts are added to distinguish house type (k) and income group (w). (Note here the way different subsystem models are linked: the income derived from jobs is related to the ability to pay for houses and so is the basis for residential 'social' differentiation.) The main equations are

$$T_{ij}^{kw} = B_j^w E_j^w V_i^{kw} e^{-\mu^w c_{ij}} \tag{10.76}$$

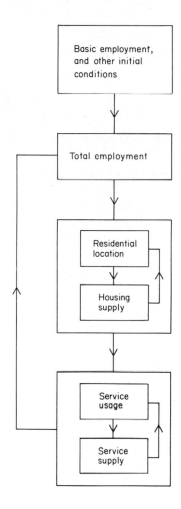

Figure 10.10 Computation scheme for an extended Lowry model.

where

$$B_j^w = \frac{1}{\sum_{ik} V_i^{kw} e^{-\mu^w c_{ij}}} \qquad (10.77)$$

to ensure that

$$\sum_{ik} T_{ij}^{kw} = E_j^w. \qquad (10.78)$$

Since the model is singly constrained, by summing $\{T_{ij}^{kw}\}$ appropriately it generates the spatial distribution of population by income:

$$P_i^w = \sum_{jk} T_{ij}^{kw}. \qquad (10.79)$$

Table 10.1

E_j^w employment in zone j in wage group w

E_i^g employment in zone j in service sector g

E_j^B employment in zone j in the basic sector

y^{wg} proportion of g-jobs offering wage w

y^{Bg} proportion of B-jobs offering wage w

T_{ij}^{kw} workers who live in a type k house in zone i and have a w-wage job in zone j

B_j^w balancing factors

V_i^{kw} the attractiveness of a type k house in zone i to a w-wage earner

μ^w residential model 'distance' parameter

c_{ij} travel (generalized) cost between i and j

P_i^w number of w-wage workers resident in zone i

b_i^{kw} the benefits for a w-wage worker associated with living in a type k house in zone i

H_i^k number of type k houses in zone i

X_{il}^k attributes of type k housing in i which determine attractiveness

λ^w expected expenditure per w-wage household on housing

q_i^k 'cost' of a type k house in zone i

ρ^k a parameter measuring the speed of system response to housing supply–demand imbalance

$e^{(1)gw}$ average per capita expenditure on service g by w-wage residents of zone i

$e^{(2)gw}$ average per capita expenditure on service g by w-wage workers in zone i

S_{ij}^{gh} flow of expenditure from zone i to zone j on service $g : h = 1$ from residents, $h = 2$ from workers

A_i^{gh} balancing factors

W_j^g attractiveness of service g in zone j, measured in terms of size of provision

α^g parameter measuring consumers' scale economies

β^{gh} 'distance' parameter for trips to service g for residence ($h = 1$) or work ($h = 2$)

D_j^g total 'revenue' attracted to service g

ε^g parameter measuring the speed of system response to supply–demand imbalance in provision of g

k_j^g cost per unit of providing g-facilities in j

γ^g number of employees per unit of size in service sector g

As discussed in Section 4.6, the form of the attractiveness function V_i^{kw} is likely to be complicated. It can, formally, be taken as an expression of preferences and if it is written in the form $e^{\mu^w b_i^{kw}}$, then the last two terms of (10.78) combine as

$$V_i^{kw} e^{-\mu^w c_{ij}} = e^{\mu^w (b_i^{kw} - c_{ij})} \tag{10.80}$$

and b_i^{kw} can be interpreted as a measure of benefits associated with the purchase of a type k house in i by a w-income household. (We are implicitly assuming one worker per household, but this can easily be relaxed.) For the present, we simply note that V_i^{kw} will be a function of a range of variables, including housing supply, H_i^k and a set of variables we write as X_{il}^k, $l = 1, 2, \ldots$ which will represent such things as density, social class mix, and so on. So, formally, we write it as

$$V_i^{kw} = V_i^{kw}(H_i^k, X_{i1}^k, X_{i2}^k, \ldots). \qquad (10.81)$$

The next step is to introduce a *model of housing supply*, H_i^k. We use the same kinds of methods as for constructing a differential equation for W_j in Section 8.4.2 (equation 8.39). Only a slight modification is needed:

$$\dot{H}_i^k = \rho^k (\sum_{jw} \lambda^w T_{ij}^{kw} - q_i^k(H_i^k, X_{il}^k)H_i^k). \qquad (10.82)$$

Here, λ^w measures the expected expenditure per household on housing and q_i^k the cost of a type k house, measured in comparable units. q_i^k is shown as a function of H_i^k and the other variables, X_{il}^k, which characterize the area. Generally, we would expect q_i^k to increase with increasing quality of these variables, and to increase very sharply as H_i^k increases beyond a threshold which represents some 'capacity'.

The equilibrium condition is

$$\sum_{jw} \lambda^w T_{ij}^{kw} = q_i^k(H_i^k, X_{il}^k)H_i^k. \qquad (10.83)$$

Equations (10.76), (10.77), (10.81), and (10.83) then constitute a complete model of both demand and supply sides which can be solved simultaneously for T_{ij}^{kn} and H_i^k.

Now consider the *service sector*. One feature of Lowry's original model, which has rarely been pursued since, was the distinction between trips made to service facilities from a residential base and those from an employment base. We retain this distinction here, using a superscript h as 1 or 2 for residence or work base, and thus taking S_{ij}^{gh} as our main variable. The base for $h = 1$ trips will be P_i^* and for $h = 2$ trips, E_i^*, where

$$P_i^* = \sum_w P_i^w \qquad (10.84)$$

and

$$E_i^* = \sum_w E_i^w \qquad (10.85)$$

and we already know the right-hand sides of each equation.

For convenience, define

$$Y_i^1 = P_i^* \qquad (10.86)$$

and

$$Y_i^2 = E_i^* \qquad (10.87)$$

and write average per capita expenditure on service g by (i) residents of i. (ii) workers in i as

$$e_i^{*g1} = \left[\sum_w e^{(1)gw} P_i^w \right] \Big/ \sum_w P_i^w \qquad (10.88)$$

$$e_i^{*g2} = \left[\sum_w e^{(2)gw} E_i^w \right] \Big/ \sum_w E_i^w \qquad (10.89)$$

where $e^{(1)gw}$ and $e^{(2)gw}$ are given aggregate constraints. The basic model equations can then be written

$$S_{ij}^{gh} = A_i^{gh} e_i^{gh} Y_i^h (W_j^g)^{\alpha^g} e^{-\beta^{gh} c_{ij}} \qquad (10.90)$$

where

$$A_i^{gh} = \frac{1}{\sum_j (W_j^g)^{\alpha^g} e^{-\beta^{gh} c_{ij}}} \qquad (10.91)$$

to ensure that

$$\sum_j S_{ij}^{gh} = e_i^{gh} Y_i^g. \qquad (10.92)$$

By summing 'the other way', we can obtain total revenue attached to service g at j:

$$D_j^g = \sum_{ih} S_{ij}^{gh}. \qquad (10.93)$$

We will assume the attractiveness function is measured by the space devoted to service g at j. More complicated assumptions are possible, but we will not pursue them here.

The supply side dynamic model—using an obvious extension of equation (8.39)—is:

$$\dot{W}_j^g = \varepsilon^g (D_j^g - k_j^g W_j^g). \qquad (10.94)$$

Here, we have taken k_j^g as a constant, but it could be made a function of other variables as in the case of q_i^k. The equilibrium conditions are obviously

$$D_j^g = k_j^g W_j^g \qquad (10.95)$$

which is a disaggregated version of equations (10.72).

Equations (10.86)–(10.95) then constitute a complete model (excluding equation (10.94) which is not needed for an equilibrium analysis) which can be solved for S_{ij}^{gh} and W_j^g.

To complete the model, we add an equation for *service employment*:

$$E_j^g = \gamma^g W_j^g. \qquad (10.96)$$

The whole set of model equations can then be solved iteratively, with the feedback from (10.96) to (10.75).

A thoughtful exploration of this set of equations shows that they can reproduce most of the phenomena which experience suggests exists.

Economic development would be built into the model by letting the E_j^B's, the Y^{gw}'s for higher income groups, and the e_i^{gh}'s all grow—to reflect expanding job opportunities at higher incomes and a corresponding increased demand for services. The urban population would grow as a consequence and new houses would be built, mainly on vacant land, and hence this fuels the decentralization process. It also could accelerate social differentiation because while housing quality will improve with increasing incomes, there will be spatial clustering. This differentiation is likely to be reinforced by terms within V_i^{kw} which express the preferences of higher income groups to associate with each other.

A hierarchical settlement structure will develop through the scale economies implied by the $(W_j^g)^{x^g}$ term in equation (10.90) and in this sense the model also functions as an alternative approach to central place theory.

The description of the model so far implies a calculation and analysis of a succession of equilibrium points—essentially a comparative static analysis. It could be argued that the structural variables, H_i^k and W_j^g, at least, move into equilibrium more slowly than the activity variable. In this case the differential equations (10.82) and (10.94)—probably in the form of difference equations— can be used directly and more complicated dynamic behaviour simulated. This would leave the possibility of the system exhibiting complicated bifurcation behaviour, including possible oscillations in some circumstances, of the types described in a more general context in Chapter 8.

We are now in a position to consider the use of models of this type, and simpler ones, in planning and we proceed to consider this topic in the next section.

10.4.5 The use of models in planning

First, recall the three different kinds of planning activities: analysis, design, and policy. Models can play a role in each of these in different ways.

The traditional role of models has been in analysis in their use as a conditional forecasting device. A 'plan' or 'design' is, in these terms, a specification of variables and parameters which are exogenous to the model. Thus the model can be used to compute the endogenous variables which is thus a prediction of the impact of the plan. We saw an example of this in Section 4.7 in relation to the basic shopping model. A set of $\{W_j^g\}$ is a plan and the model can be used to calculate $\{D_j^g\}$, the set of revenues, which is a measure of the impact of shopping centre developers. It is also possible to calculate the impact in terms of consumers' surplus (Section 6.2.7), and even to consider a struggle between the two interests with the state as mediator (the example of comparative static analysis in Section 8.2.1).

The Lowry-type model of the previous section could be used in the same way, and the larger set of endogenous variables would provide a more comprehensive assessment of impacts, though there would also be a need for more exogenous information.

Two other uses of models in analysis for planning can usefully be distinguished. First, they can be used to analyse, and perhaps to anticipate, problems. Earlier, for example, we mentioned the problem of increasing energy prices; such increases could be reflected in parameters like β^{wh} and μ^w—i.e. they would increase as trip lengths shortened. But the model would reveal other changes in the settlement structure. The consequences of increasing unemployment could be similarly investigated, as could the changing pattern of social polarization.

Secondly, the methods of Chapter 8 could be used to investigate the stability of structures and, ideally, the critical values of parameters or exogenous variables at which states become unstable. A concern with stability and criticality — and what Holling calls 'resilience' for ecosystems — could turn out to be a more important contribution of models in planning than conditional forecasting, if only because the latter always has the weakness associated with the inevitable changes to come.

The contribution of models to design is probably mainly to be made through the assessment of impacts following the 'manual' manipulation of controllable variables. When this method is used in connection with an interactive computer system, it can be a very powerful tool. But models of the type described in the previous subsection — or simpler ones as we saw in the case of the shopping model in Sections 4.3 and 8.2.1 — can be embedded in an overall optimization framework, for example to maximize consumers' surplus. In effect, this determines an optimal allocation of basic-sector jobs or shopping centres, or whatever, but with the inbuilt assumption that consumers will continue to behave in the same way as implied by the analytical models. Even if the procedures are imperfect, they may generate a useful first step in generating designs.

Finally, the models have a direct analytical contribution to policy making through the construction of evaluation indicators which can then be used within a goals–achievement matrix, cost–benefit analysis, or whatever. These can be aggregative — change in consumers' surplus being an obvious example — or much more disaggregated, like measures of accessibility for certain types of residents at each location to particular types of facilities (as described in Section 4.6). The detailed prediction of impacts is, of course, useful as well and the kinds of indicators mentioned are, in effect, summaries of these. It can be argued that, in this sense, models have a crucial role to play at Beer's level 4 of an urban planning system. They are the heart of the intelligence system; they generate 'variety' which can be used in the control system. And, further, the information should be made publicly available (i.e. to Beer's 'environment') and this should aid public control through the various institutions which relate to the planning system itself.

10.4.6 Future developments in urban modelling

It should be clear even from the relatively brief discussion in this book that urban models have a lot to offer in different kinds of urban planning. It is important to have a good idea of the future demand for a wide variety of services. Two current examples are falling school rolls resulting from a declining birthrate, and increasing demands for health services from an increasing number of older people in the population. This also illustrates an important general point: the time has perhaps come when the urban modelling fields should break away from its particular associations with *town planning* and be seen to play a much wider role in urban government. We take up this point in more detail below.

The other models used also have specific uses. The economic model, for example, could be used to represent changing unemployment patterns or as a basis for forecasting the demand for water or energy. The interaction models predict demands for or allocations to housing or various services.

However, the account of the state-of-the-art given above also indicates that the field has a long way to develop. Many of the characteristics of system elements listed in Section 10.4.2 do not appear in models and even more appear only in small subsets in subsystems models, and many interdependencies are thus lost. This is one reason for the development of micro-simulation models as described in Chapter 8, and such attempts to incorporate more information and more relationships in models may be an important feature of future work. But also, we have seen that real life is likely to be much more complicated than current models in the complexity of its dynamics: lags and a variety of functional forms within differential or difference equations can generate an immense range of possible system behaviours and these have yet to be explored more fully.

There are corresponding problems in the planning side, though possibly at a more basic level. We observed above that urban models have much potential for planning now. The fact is that they are used far less than they could be. This is probably a statement about the education of planners: first, there are not enough planners available with the appropriate skills to use modelling techniques; secondly, there is insufficient awareness of the need for greater analytical capability within the planning process. Both of these feature came together in Beer's level-4 intelligence system. In most planning organizations, no attempt has been made to construct such a system on a sufficiently substantial scale (which means with adequate variety, using models to help to generate this variety). One conclusion for the future, therefore, is that intelligence systems — which can be seen as data banks or information systems improved by the use of associated models — should be established on a bigger scale, possibly with central government guidance. These could then form the foundation of a more sophisticated planning system. Such a system could be more efficient internally and could also lead to more equitable distributions of resources and more effective democratic institutions through the information being shared with the public at large.

Finally, it is useful to make distinctions between research, development, and application in assessing the future of models in urban planning. Typically more people are involved in development activities than research, and more in applications than development. We have argued, implicitly, that this distribution of manpower does not currently satisfy this condition. Nor, indeed, is there the implied division of labour: quite often, the same people are trying to carry out two or three of what are rather different kinds of function. In this book, we have concentrated on the presentation of a wide range of methods which can be applied in urban modelling — and elsewhere, of course. In our examples, we have presented models which are usable immediately, but have also tried to chart the research frontier. This should provide a framework within which the reader can assess the possible forms of future activity: increasing volumes of work of development and application kind with currently usable models coupled with a research programme on some very difficult new problems.

10.5 CONCLUDING COMMENTS

At various stages in this book, we have characterized systems analysis as being concerned with (i) handling complexity, (ii) identifying and understanding systemic effects — for example, so that system behaviour is no longer counter-intuitive, (iii) seeking methods which are applicable to a wide range of systems, grouped into a number of 'types', and (iv) providing tools as an aid to planning and problem solving. At the heart of this programme lies the task of building models of systems and subsystems which provide the analytical basis of understanding and of planning. What progress has been made?

In addressing this question, we should perhaps first note that the application of systems analytical techniques to the environment is a relatively young field. It has developed only within the last twenty-five years or so on any substantial scale, and in many instances only much more recently than that. In this sense, progress has been very rapid: there are at least rudimentary models available, for instance, in relation to the three system-examples we have used throughout the book. And with the arrival of a range of techniques for dynamic modelling and the expansion of optimization techniques, it could be argued that most of the appropriate methods, again at least in a rudimentary form, are now available. Indeed, there are many alternative methods and models available in many particular instances. What is perhaps most lacking in many cases is an adequate level of empirical work which would lead to further refinement of theoretical models and to an ability to choose between alternatives. In some fields, such as transport modelling and planning, there has been an immense amount of empirical work, but this is very much the exception rather than the rule.

We noted at the end of the previous section that the urban systems field should be examined in relation to the concepts of research, development, and application and the way effort is organized in this sense. The same comments apply more broadly. One of the reasons for lack of empirical work is the failure to apply systems analytical techniques on a large scale in organizations at the 'applications end' of the spectrum. These organizations are usually large enough to have the capability in principle and, in particular, to establish appropriate data bases. However, before there is a major change in this respect, more effort is needed in development. This is likely to be the responsibility of central government — either Ministries directly or associated scientific laboratories. The time may come when a greater better-directed and more-sustained effort is launched. Indeed, at a time when resources are seen to be scarce, and when serious scrutiny needs to be given both to efficiency on the one hand and to equity and democracy on the other — both of which need the best kind of information base — then systems analysis, poised with the right kinds of techniques, could come into its own.

There is a general argument, therefore, for some optimism about the progress which can be made if the associated organizational structures are right. In terms of the objectives of systems analysis we can note in relation to our examples that a wide range of tools does exist to enable us to handle complexity. We have identified systemic effects and feel that the techniques available would enable us to attack any new problems of 'counter-intuitive' behaviour. The tool-kit of Chapters 4–9 is not yet fully comprehensive, but does take us some way towards the third objective of being able to assemble models for any particular system — hard though the associated research problems might be. And we have seen a number of examples, and citied others, of the usefulness of these techniques in a planning context. Finally, although we have restricted the presentation of the argument to three illustrative examples, it is easy to see that the same kinds of methods are more widely applicable to environmental systems. So we can conclude on a note which is optimistic in principle while recognizing that there are many research problems for the future. The application of systems analysis to the spatial and environmental sciences will continue to be an exciting field in which to work.

NOTES ON FURTHER READING

The methods used to build the moorland ecological model in Section 10.2 are based on these of *Maynard Smith (1974, Chapter 10). Otherwise, the references cited at the end of Chapter 1 suffice.

The water resources examples are taken from †Biswas (1976) and †de Neufville and Marks (1974). The main authors used in these books were Loucks (two papers) and Howe from Biswas, and de Neufville, Bishop and Hendricks, Jacoby and Loucks, and Leclerc and Marks from de Neufville and Marks. We should perhaps also add that the classic original text is that of *Maas et al. (1962).

The urban systems model which provides the focal point of Section 10.4 on urban systems is developed from the in *Wilson (1981, Chapter 6). A more detailed account will be found in that book together with comparisons with alternative approaches. A more general survey of urban models is to be found in †Wilson (1974).

Bibliography

Ackerman, E. (1958). Geography as a fundamental research discipline. *Research Paper 33*, Department of Geography, University of Chicago.

Ackerman, E. (1963). Where is a research frontier? *Annals, Association of American Geographers*, **53**, 429–40.

Alexander, C. (1965). *Notes on the synthesis of form*, Harvard University Press, Cambridge, Mass.

Allen, P. M., Deneubourg, J. L., Sanglier, M., Boon, F., and de Palma, A. (1978). *The dynamics of urban evolution*, Vol. 1: Interurban evolution; Vol. 2: Intraurban evolution. Final Report to the U.S. Department of Transportation, Washington D.C.

Artle, R. (1959). *Studies in the structure of the Stockholm economy*, Business Research Institute, Stockholm School of Economics. Republished, 1965: Cornell University Press, Ithaca.

Ashby, W. R. (1956). *An introduction to cybernetics*, Chapman and Hall, London.

Atkin, R. H. (1974). *Mathematical structure in human affairs*, Heinemann, London.

Atkin, R. H. (1977). *Combinatorial connectiveness in social systems*, Birkhauser, Basel.

Balinski, M. L. and Baumol, W. J. (1968). The dual non-linear program and its economic interpretations. *Review of Economic Studies*, **35**, 237–56.

Barrows, H. H. (1923). Geography as human ecology. *Annals, Association of American Geographers*, **13**, 1–14.

Bartholomew, D. J. (1967; 2nd edn, 1974). *Stochastic models for social processes*, Wiley, Chichester.

Batty, M. (1976). *Urban modelling: algorithms, calibration, predictions*, Cambridge University Press, Cambridge.

Beaumont, J. R., Clarke, M., and Wilson, A. G. (1980). The dynamics of urban spatial structure; some exploratory results using difference equations and bifurcation theory. *Proceedings of Cambridge Conference 11th–13th September 1980*, forthcoming.

Beer, S. (1967) *Management Science*, Aldus Books, London.

Beer, S. (1972; Second Edition, 1980a). *Brain of the firm*, Allen Lane, The Penguin Press, Harmondsworth. 2nd edn: Wiley Chichester.

Beer, S. (1980b). *The heart of enterprise*, Wiley, Chichester.

Bennett, R. J. and Chorley, R. J. (1978). *Environmental systems: philosophy, analysis and control*, Methuen, London.

Berlinsky, D. (1976). *On systems analysis*, M.I.T. Press, Cambridge, Mass.

Bertalanffy, L. von (1968). *General system theory*, Braziller, New York.

Bishop, A. B. and Hendricks, D. W. (1974). Analysis of water re-use alternatives in an integrated urban and agricultural area. In R. de Neufville and D. H. Marks (Eds), *Systems planning and design*, Prentice Hall, Englewood Cliff. pp. 169–89.

278

Biswas, A. K. (Ed.) (1976). *Systems approach to water management*, McGraw-Hill, New York.

Boyce, D. E., Fahri, A., and Weischedel, R. (1974). Optimal network problem: a branch and bound algorithm. *Environment and Planning*, **5**, 519–33.

Burton, I. (1963). The quantitative revolution and theoretical geography. *Canadian Geographer*, 7, 151–62.

Carruthers, G. A. P. (1956). An historical review of the gravity and potential concepts of human interaction. *Journal of the American Institute of Planners*, **22**, 94–100.

Chadwick, G. (1970; 2nd edn, 1978). *A systems view of planning*, Pergamon, Oxford.

Chapman, G. T. (1977). *Human and environmental systems*, Academic Press, London.

Chilton, R. and Poet, R. R. W. (1973). An entropy maximising approach to the recovery of detailed migration patterns from aggregate census data. *Environment and Planning*, **5**, 135–46.

Chisholm, M. (1967). General systems theory and geography. *Transactions, Institute of British Geographers*, **42**, 42–52.

Chorley, R. J. and Kennedy, B. A. (1971). *Physical geography: a systems approach*, Prentice Hall, Englewood Cliffs.

Clarke, M. (1977). The development of an environmental simulation game. *Working Paper 208*, School of Geography, University of Leeds.

Clarke, M. (1978). Corporate planning in local government. A review and further developments towards a systematic modelling framework. *Working Paper 221*, School of Geography, University of Leeds.

Clarke, M., Keys, P., and Williams, H. C. W. L. (1981). Micro analysis and simulation of socio-economic systems: progress and prospects. In Bennett, R. J. and Wrigley, N. (Eds), *Quantitative Geography in Britain, retrospect and prospect*, Routledge and Kegan Paul, London.

Cripps, E. L., Macgill, S. M., and Wilson, A. G. (1974). Energy and materials flows in the urban space economy. *Transportation Research*, **8**, 293–305.

Department of Health and Social Security (1980). *Patients first*. Consultative paper on the structure and management of the National Health Service in England and Wales, HMSO, London.

Eddison, T. (1973). *Local government: management and corporate planning*, Leonard Hill Books, Aylesbury.

Evans, S. P. (1973). A relationship between the gravity model for trip distribution and the transportation problem of linear programming. *Transportation Research*, 7, 39–61.

Evans, S. P. (1976). Derivation and analysis of some models for combining trip distribution and assignment. *Transportation Research*, **10**, 37–57.

FitzPatrick, E. A. (1974). *An introduction to soil science*, Oliver and Boyd, Edinburgh.

Forrester, J. W. (1968). *Principles of systems*, Wright-Allen Press, Boston.

Forrester, J. W. (1969). *Urban dynamics*, M.I.T. Press, Cambridge, Mass.

Gimingham, C. H. (1972). *Ecology of heathlands*, Chapman and Hall, London.

Gimingham, C. H. (1975). *An introduction to heathland ecology*, Oliver and Boyd, Edinburgh.

Gould, P. R. (1972). Pedagogic review: entropy in urban and regional modelling. *Annals, Association of American Geographers*, **62**, 689–700.

Greenwood, R. and Stewart, J. D. (Eds) (1975). *Corporate planning in English local government*, Charles Knight, London.

Gulick, L. H. and Urwick, L. (1937). *Papers on the science of administration*, Institute of Public Administration, New York.

Haggett, P. (1965). *Locational analysis in human geography*, Edward Arnold, London.

Haggett, P. and Chorley, R. J. (1969). *Network analysis in geography*, Edward Arnold, London.

Haggett, P., Cliff, A. D., and Frey, A. (1977). *Locational analysis in human geography*, 2nd edn, Edward Arnold, London.

280

Hall, A. D. and Fagen, R. E. (1956). Definition of a system. *General Systems Yearbook*, **1**, 18–28.
Hall, P. (1974). *Urban and regional planning*, Penguin, Harmondsworth.
Hall, P. (1980). *Great planning disasters*, Weidenfeld and Nicholson, London.
Hansen, W. G. (1959). How accessibility shapes land use. *Journal of the American Institute of Planners*, **25**, 73–6.
Harris, B. (1965). Urban development models: a new tool for planners. *Journal of the American Institute of Planners*, **32**, 90–5.
Harris, B. (1981). *A paradigm for planning*, University of California Press, Berkeley.
Hay, A. (1973). *Transport for the space economy: a geographical study*, Macmillans, London.
Hirsch, M. W. and Smale, S. (1974). *Differential equations, dynamical systems, and linear algebra*, Academic Press, New York.
Howe, C. W. (1976). Economic models. In A. K. Biswas (Ed.), *Systems approach to water management*, McGraw Hill. New York. pp. 335–64.
Hudson, L. (1980). Language, truth and psychology. In L. Micheals and C. Ricks (Eds.), *The state of the language*, University of California Press, Berkeley. pp. 449–57.
Huff, D. L. (1964). Defining and estimating a trading area. *Journal of Marketing*, **28**, 34–8.
Huggett, R. (1980). *Systems analysis in geography*, Oxford University Press, Oxford.
Isard, W., et al. (1972). *Ecologic–economic analysis for regional development*, The Free Press, New York.
Jacoby, H. D. and Loucks, D. P. (1974). The combined use of optimisation and simulation models in river basin planning. In R. de Neufville and D. H. Marks (Eds.), *Systems planning and design*, Prentice Hall, Englewood Cliffs. pp. 237–56.
Jaynes, E. T. (1957). Information theory and statistical mechanics. *Physical Review*, **106**, 620–30.
Jeffers, J. N. R. (1978). *An introduction to systems analysis: with ecological applications*, Edward Arnold, London.
Kelly, R. A. and Spofford, W. O. (1977). Application of an ecosystem model to water quality management: the Delaware estuary. In C. A. S. Hall and J. W. Day (Eds), *Ecosystem modelling in theory and practice*, Wiley, New York.
King, L. J. and Golledge, R. G. (1978). *Cities, space and behaviour: the elements of urban geography*, Prentice Hall, Englewood Cliffs.
Klir, J. and Valach, M. (1967). *Cybernetic modelling*, Iliffe, London.
Kuhn, T. (1962; Second Edition, 1980). *The structure of scientific revolutions*, University of Chicago, Chicago.
Kuhn, T. (1963). The essential tension: tradition and innovations in scientific research. In C. W. Taylor and F. Barron (Eds), *Scientific creativity: its recognition and development*, Wiley, New York. Also reprinted in L. Hudson (Ed.) (1970), *The ecology of human intelligence*, Penguin, Harmondsworth.
Kullback, S. (1959). *Information theory and statistics*, Wiley, New York.
Lakshmanan, T. R., and Hansen, W. G. (1965). A retail market potential model. *Journal of the American Institute of Planners*, **31**, 134–43.
Langbein, W. B. and Leopold, L. B. (1964). Quasi-equilibrium states in channel morphology. *American Journal of Science*, **262**, 782–94.
Leclerc, G. and Marks, D. H. (1974). Determination of the discharge policy for existing reservoir networks. In R. de Neufville and D. H. Marks (Eds), *Systems planning and design*, Prentice Hall, Englewood Cliffs. pp. 257–72.
Leontief, W. W. (1951). Input–output economics. *Scientific American*, **151**, 15–21.
Leontief, W. W. (1967). *Input–output analysis*, Oxford University Press, Oxford.
Leslie, P. H. (1945). On the use of matrices in certain population mathematics. *Biometrika*, **23**, 183–212.

281

Leavis, F. R. (1962). *Two cultures? The significance of C. P. Snow*, Chatto and Windus, London.

Lichfield, N. (1969). Cost–benefit analysis in urban expansion: a case study, Peterborough. *Regional Studies*, **3**, 123–55.

Lichfield, N. (1970). Evaluation methodology of urban and regional plans: a review. *Regional Studies*, **4**, 151–65.

Lowry, I. S. (1964). *A model of metropolis*, RM-4035-RC, Rand Corporation, Santa Monica.

Loucks, D. P. (1976a). Surface-water quantity management models. In A. K. Biswas (Ed.), *Systems approach to water management*, McGraw Hill, New York. pp. 156–218.

Loucks, D. P. (1976b). Surface-water quality management models. In A. K. Biswas (Ed.), *Systems approach to water management*, McGraw Hill, New York. pp. 219–252.

Luce, R. D. (1959). *Individual choice behaviour*, Wiley, New York.

Maass, A., et al. (1962). *Design of water resource systems*, Harward University Press, Cambridge, Mass.

Macgill, S. M. (1975). Balancing factor methods in urban and regional analysis. *Working Paper 124*, School of Geography, University of Leeds.

Macgill, S. M. (1977a). Rectangular input–output tables—multiplier analysis and entropy maximising principles: a new methodology. *Regional Science and Urban Economics*, **8**, 355–70.

Macgill, S. M. (1977b). Theoretical properties of biproportional matrix adjustments. *Environment and Planning, A*, **9**, 687–701.

MacKinnon, R. D. and Hodgson, M. J. (1969). The highway system of southern Ontario and Quebec: some simple network generation models. *Research Report 18*, Centre for Urban and Community Studies, University of Toronto.

McLoughlin, J. B. (1969). *Urban and regional planning: a systems approach*, Faber, London.

McLoughlin, J. B. (1973). *Control and urban planning*, Faber, London.

Marglin, S. A. (1963). *Dynamic investment planning*, Harvard University Press, Cambridge, Mass.

Martin, R. L., Thrift, N. J., and Bennett, R. J. (Eds) (1979). *Towards the dynamic analysis of spatial systems*, Pion, London.

Maruyama, M. (1963). The second cybernetics: deviation-amplifying mutual causal processes, *American Scientist*, **51**, 164–79.

May, R. M. (1976). Simple mathematical models with very complicated dynamics. *Nature*, **261**, 459–67.

Maynard Smith, J. (1974). *Models in ecology*, Cambridge University Press, Cambridge.

Medawar, P. B. (1969). *Induction and intuition in scientific thought*, Methuen, London.

Miliband, R. (1976). *Marxism and politics*, Oxford University Press, Oxford.

Miller, J. G. (1978). *Living systems*, McGraw-Hill, New York.

Miller, R. E. (1979). *Dynamic optimisation and economic applications*, McGraw-Hill, New York.

Neufville, de R. (1974). Systems analysis of large-scale public facilities: New York City's water supply network as a case study. In R. de Neufville and D. H. Marks (Eds), *Systems planning and design*, Prentice Hall, Englewood Cliffs. pp. 30–47.

Neufville, de R. and Marks, D. H. (Eds) (1974). *Systems planning and design*, Prentice Hall, Englewood Cliffs.

Nystuen, J. D. and Dacey, M. F. (1961). A graph theory interpretation of nodal regions. *Papers, Regional Science Association*, **7**, 29–42.

Odum, E. P. (1975). *Ecology*, Second Edition, Holt, Rinehart, and Winston, London and New York.

Olsson, G. (1965). *Distance and human interaction*, Regional Science Research Institute, Philadelphia.

Openshaw, S. (1975). *Some theoretical and applied aspects of spatial interaction shopping models*, Geo-Abstracts, Norwich.

Orcutt, G. H., *et al.* (1961). *Micro-analysis of socio-economic systems*, Harper, New York.

O'Sullivan, P., Holtzclaw, G. D., and Barber, G. (1979). *Transport network planning.* Croom Helm, London.

Patten, B. C. (1971). A primer for ecological modelling and simulation with analog and digital computers. In B. C. Patten (Ed.), *Systems analysis and simulation in ecology*, Academic Press, New York. pp. 3–121.

Pielou, E. C. (1969). *An introduction to mathematical ecology*, Wiley, New York.

Popper, K. R. (1959). *The logic of scientific discovery*, Hutchinson, London.

Poston, T. and Stewart, I. (1978). *Catastrophe theory and its applications*, Pitman, London.

Potts, R. B. and Oliver, R. M. (1972). *Flows in transportation networks*, Academic Press, New York.

Quade, E., Brown, K., Levien, R., Majone, G., and Rakhmankulov, V. (1976). *Systems analysis: an outline for the state-of-the-art survey publications*, RR-76-16, IIASA, Laxenburg.

Rees, P. H. and Wilson, A. G. (1977). *Spatial population analysis*, Edward Arnold, London.

Revelle, C. S. (1968). Central facilities location. *Report 1002*, Centre for Environmental Quality Management, Cornell University.

Robert, F. S. (1976). *Discrete mathematical models*, Prentice Hall, Englewood Cliffs.

Rogers, A. (1966). Matrix methods of population analysis. *Journal of the American Institute of Planners*, **32**, 40–4.

Rogers, A. (1971). *Matrix methods in urban and regional analysis*, Holden Day, New York.

Rogers, A. (1975). *Introduction to multiregional mathematical demography*, Wiley, New York.

Sayer, R. A. (1976). *A critique of urban modelling: from regional science to urban and regional political economy*, Pergamon, Oxford.

Scott, A. J. (1971). *Combinatorial programming, spatial analysis and planning*, Methuen, London.

Scott, A. J. and Roweis, S. (1977). Urban planning in theory and practice: a re-appraisal. *Environment and Planning, A*, **9**, 1097–1119.

Self, P. (1972). *Administrative theories and politics: an inquiry into the structure and processes of modern government*, Allen and Unwin, London.

Senior, M. L. (1979). From gravity modelling to entropy maximising: a pedogogic guide. *Progress in Human Geography*, **3**, 179–210.

Shannon, C. and Weaver, W. (1949). *The mathematical theory of communication*, University of Illinois Press, Urbana.

Smith, R. T. (1977a). Bracken in Britain I: background to the problem of bracken infestation. *Working Paper 189*, School of Geography, University of Leeds.

Smith, R. T. (1977b). Bracken in Britain II: ecological observations of a bracken population over a six year period. *Working Paper 190,* School of Geography, University of Leeds.

Smith, R. T. (1977c). Bracken in Britain III: towards alternative strategies of bracken control with the use of herbicide. *Working Paper 191,* School of Geography, University of Leeds.

Snow, C. P. (1959). *The two cultures and the scientific revolution*, Cambridge University Press, Cambridge.

Stearns, F. W. and Montag, J. (Eds) (1975). *The urban ecosystem*, Halstead Press, New York; Wiley, Chichester.

Steenbrink, P. A. (1974). *Optimization of transport networks*, Wiley, Chichester.

Steger, W. (1965). A review of analytical techniques for the C.R.P. *Journal of the American Institute of Planners*, **31**, 166–72.

Stewart, J. D. (1971). *Management in local government*, Charles Knight, London.

Stewart, J. D. (1974). *The responsive local authority*, Charles Knight, London.

Stoddart, D. (1965). Geography and the ecological approach: the ecosystem as a geographic principle and method. *Geography*, **50**, 242.

Stone, R. (1966). *Mathematics in the social sciences*, Chapman and Hall, London.

Stone, R. (1970). *Mathematical models of the economy*, Chapman and Hall, London.

Strahler, A. W. (1965). *Introduction to physical geography*, Wiley, New York.

Thom, R. (1975). *Structural stability and morphogenesis*, W. A. Benjamin, Reading, Mass.

Vester, F. (1976). *Urban systems in crisis*, Deutschen Verlags-Anstalt, Stuttgart.

Ward, R. C. (1973). *Principles of hydrology*, 2nd edn, McGraw-Hill, London.

Wardrop, J. C. (1952). Some theoretical aspects of road traffic research. *Proceedings of the Institution of Civil Engineers, Part II*, **1**, 325–78.

Weaver, W. (1958). A quarter century in the natural sciences. *Annual Report, The Rockefeller Foundation, New York*, 7–122.

Webber, M. J. (1979). *Information theory and urban spatial structure*, Croom Helm, London.

Williams, H. C. W. L. (1977). On the formation of travel demand models and economic evaluation measures of user benefit. *Environment and Planning, A*, **9**, 285–344.

Wilson, A. G. (1968). Models in urban planning: a synoptic review of recent literature. *Urban Studies*, **5**, 249–76. Reprinted in Wilson (1972c), *Papers in urban and regional analysis*, Pion, London.

Wilson, A. G. (1970). *Entropy in urban and regional modelling*, Pion, London.

Wilson, A. G. (1971). A family of spatial interaction models and associated developments. *Environment and Planning*, **3**, 1–32. Reprinted in Wilson (1972c), *Papers in urban and regional analysis*, Pion, London.

Wilson, A. G. (1972a). Theoretical geography: some speculations. *Transactions, Institute of British Geographers*, **57**, 31–44.

Wilson, A. G. (1972b). Understanding the city of the future. *University of Leeds Review*, **15**, 135–66.

Wilson, A. G. (1972c). *Papers in urban and regional analysis*, Pion, London.

Wilson, A. G. (1973a). How planning can face new issues. In P. Cowan, (Ed.), *The future of planning*, Heinemann, London. pp. 24–43.

Wilson, A. G. (1973b). Towards system models for water resource management. *Journal of Environmental Management*, **1**, 36–52.

Wilson, A. G. (1974). *Urban and regional models in geography and planning*, Wiley, Chichester and New York.

Wilson, A. G. (1981). *Catastrophe theory and bifurcation: applications to urban and regional systems*, Croom Helm, London, University of California Press, Berkeley.

Wilson, A. G., Coelho, J. D., Macgill, S. M., and Williams, H. C. W. L. (1981). *Optimisation in locational and transport analysis*, Wiley, Chichester.

Wilson, A. G. and Kirkby, M. J. (1975; 2nd edn, 1980). *Mathematics for geographers and planners*, Oxford University Press, Oxford.

Wilson, A. G. and Pownall, C. (1976). A new representation of the urban system for modelling and for the study of micro-level interdependence. *Area*, **5**, 246–54.

Wilson, A. G., Rees, P. H., and Leigh, C. M. (Eds) (1977). *Models of cities and regions*, Wiley, Chichester.

Wilson, A. G. and Senior, M. L. (1974). Some relationships between entropy maximising models, mathematical programming models and their duals. *Journal of Regional Science*, **14**, 205–15.

Woldenberg, M. J. (1970). A structural taxonomy of spatial hierarchies. In M. Chisholm, A. E. Frey, and P. Haggett, (Eds), *Regional forecasting*, Butterworths, London.

Yorkshire and Humberside Economic Planning Board (1976). *The Pennine Uplands*, HMSO, London.

Zeeman, E. C. (1977). *Catastrophe theory*, Addison-Wesley, Reading, Mass.

Ziman, J. M. (1968). *Public knowledge*, Cambridge University Press, Cambridge.

Index

285